Hands-On
Visual Studio 2022

A developer's guide to exploring new features and
best practices in VS2022 for maximum productivity

Miguel Angel Teheran Garcia

Hector Uriel Perez Rojas

BIRMINGHAM—MUMBAI

Hands-On Visual Studio 2022

Copyright © 2022 Packt Publishing

Group Product Manager: Pavan Ramchandani

Publishing Product Manager: Aaron Tanna

Senior Editor: Mark Dsouza

Content Development Editor: Rashi Dubey

Technical Editor: Simran Udasi

Copy Editor: Safis Editing

Project Coordinator: Ajesh Devavaram

Proofreader: Safis Editing

Indexer: Pratik Shirodkar

Production Designer: Joshua Misquitta

Marketing Coordinator: Anamika Singh

First published: June 2022

Production reference: 1290622

Published by Packt Publishing Ltd.

Livery Place

35 Livery Street

Birmingham

B3 2PB, UK.

ISBN 978-1-80181-054-8

www.packt.com

To my mother, Micaela Garcia, and my sister, Maria Angelica Teheran, for teaching me the value of hard work, and special mention to my wife, Rina Plata, for supporting me during the evolution of my professional career and personal life.

– Miguel Angel Teheran Garcia

To God, first and foremost, for his blessings. To my wife, Cristina, for her patience and love, to my beloved children, Elizabeth and Ricardo, for their antics that make me laugh every day, to my parents Isaias and Carmen, and brothers César, Miguel and Leonardo for their unconditional support.

– Hector Uriel Perez Rojas

Contributors

About the authors

Miguel Angel Teheran Garcia is a solutions architect expert in Microsoft technologies, with over 10 years of experience. In 2020, he was recognized as a Microsoft MVP and, in 2021, as an Alibaba Cloud MVP. Miguel is an active member of tech communities and a content author with C# Corner.

"I want to thank my family, my parents, and friends that have been helping and supporting me in every new challenge in my personal life and professional career. I want to thank the Avanet community for being my close friends and who have supported me, especially when I started to work as a software developer. Finally, I also want to say thanks to all people around the world that read my articles, watch my videos, and take my courses. All the feedback provided by them makes me a better professional."

Hector Uriel Perez Rojas is an experienced senior developer, with more than 10 years of experience in developing desktop, web, and mobile solutions with the .NET platform. He was recognized in 2021 with the Microsoft MVP award. He is an active member of the community and has his own training academy.

"I want to thank my family for their unconditional support and love, the developer community for always being so active, those who have taught me something valuable throughout my life, and all the people who have made the publication of this book possible."

About the reviewer

Robin Schroeder has been writing code since 1999. She initially worked in Java and eventually made the leap to C# in 2008. She focused on building websites and Windows 8 Metro apps until 2015, when she switched over to writing Xamarin cross-platform mobile apps. She currently focuses on mobile phone BLE communications with IoT devices. She is also a technical public speaker and works with a local high school girls' coding club during the school year.

"I would like to thank my husband for his love and support. I am delighted to help out on projects that encourage young people to code.

I know this book will do just that."

Table of Contents

3

Improvements in Visual Studio 2022

4

Creating Projects and Templates

5

Debugging and Compiling Your Projects

Part 2: Tools and Productivity

6

Adding Code Snippets

7

Coding Efficiently with AI and Code Views

8

Web Tools and Hot Reload

9

Styling and Cleanup Tools

10

Publishing Projects

Part 3: GitHub Integration and Extensions

11

Implementing Git Integration

12

Sharing Code with Live Share

Preface

*Visual Studio 2022 is a complete and ideal **integrated development environment** (IDE) for creating large, complex, and scalable applications. It is one of the most complete tools available for development, especially with Microsoft technologies.*

This book will teach you how to take advantage of the tools available with this IDE to write clean code faster. You'll begin by learning how to set up and start Visual Studio 2022 and use all the tools provided by this IDE. You will then learn key combinations, tips, and additional utilities that can help you to code faster and review your code constantly. You will see how to compile, debug, and inspect your project to analyze its current behavior using Visual Studio. You will see how to insert reusable blocks of code writing simple statements. You will learn about some visual aids and artificial intelligence that will help you improve productivity and understand what is going on in a project.

By the end of this book, you will be able to set up your development environment using Visual Studio 2022, personalize the tools and layout, and use shortcuts and extensions to improve your productivity.

Who this book is for

This book is for .NET software developers focusing on web development and web developers who want to learn about the new features, tools, and features available in Visual Studio 2022. Basic knowledge of HTML, CSS, and JavaScript or frameworks such as React and Angular is assumed.

What this book covers

Chapter 1, Getting Started with Visual Studio 2022, reviews how to install Visual Studio, the versions, and requirements.

Chapter 2, Configuring the IDE, discusses how to configure Visual Studio depending on your needs and preferences.

Chapter 3, Improvements in Visual Studio 2022, reviews the new features and improvements included in Visual Studio 2022.

Chapter 4, Creating Projects and Templates, explains the templates included in Visual Studio and how to create your first project.

Chapter 5, Debugging and Compiling Your Projects, discusses how to use Visual Studio to build applications and debug your projects.

Chapter 6, Adding Code Snippets, explains how to use code snippets and create your own in Visual Studio.

Chapter 7, Coding Efficiently with AI and Code Views, reviews the AI tools and different code views included in Visual Studio.

Chapter 8, Web Tools and Hot Reload, explains the web tools included in Visual Studio for web developers and how Hot Reload works.

Chapter 9, Styling and Cleanup Tools, discusses the code cleanup options included in Visual Studio for backend and frontend developers.

Chapter 10, Publishing Projects, explores the different ways to publish projects from Visual Studio.

Chapter 11, Implementing Git Integration, reviews the Visual Studio functionalities to work with GitHub-hosted projects.

Chapter 12, Sharing Code with Live Share, discusses what Live Share is and how to use it to work with teams in live coding sessions.

Chapter 13, Working with Extensions in Visual Studio, explains what extensions are in Visual Studio and the different ways to add them to the IDE.

Chapter 14, Using Popular Extensions, discusses what the most popular extensions for Visual Studio are and why.

Chapter 15, Learning Keyboard Shortcuts, explains how to perform quick actions in Visual Studio using the keyboard to optimize repetitive tasks.

To get the most out of this book

You need to have a laptop or desktop computer with Windows 10 or later installed. To install Visual Studio and complete the exercises in Chapters 10–14, you will need an internet connection.

To obtain a Visual Studio Community license, you must have a Microsoft account, either belonging to the Hotmail domain or the Outlook domain.

Software/hardware covered in the book	Operating system requirements
A 1.8 GHz or faster 64-bit processor; quad-core or better is recommended. ARM processors are not supported. A minimum of 4 GB of RAM. Many factors impact resources used; we recommend 16 GB of RAM for typical professional solutions. Hard disk space – a minimum of 850 MB up to 210 GB of available space, depending on the features installed. Typical installations require 20–50 GB of free space.	Windows 10 or later
Visual Studio 2022 Community Edition	Windows 10 or later

A version of Visual Studio Enterprise is required to generate code maps in Chapter 8, Web Tools and Hot Reload.

To perform the tests suggested in Chapter 11, Implementing Git Integration, a GitHub account is required.

If you are using the digital version of this book, we advise you to type the code yourself or access the code from the book's GitHub repository (a link is available in the next section). Doing so will help you avoid any potential errors related to the copying and pasting of code.

To complete and understand all the activities throughout this book, it's important to have knowledge about software development and web development (HTML, JavaScript, and CSS).

Download the example code files

You can download the example code files for this book from GitHub at `https://github.com/PacktPublishing/Hands-On-Visual-Studio-2022`. If there's an update to the code, it will be updated in the GitHub repository.

We also have other code bundles from our rich catalog of books and videos available at `https://github.com/PacktPublishing/`. Check them out!

Download the color images

We also provide a PDF file that has color images of the screenshots and diagrams used in this book. You can download it here: `https://packt.link/VHA6o`.

Conventions used

There are a number of text conventions used throughout this book.

`Code in text`: Indicates code words in text, database table names, folder names, filenames, file extensions, pathnames, dummy URLs, user input, and Twitter handles. Here is an example: "The template has a demo with the `WeatherForecastController.cs` file."

A block of code is set as follows:

```
CommonMethod("Before invocation of NewMethod()");
NewMethod();
CommonMethod("After invocation of NewMethod()");
```

When we wish to draw your attention to a particular part of a code block, the relevant lines or items are set in bold:

```
public float Calculate1()
{
    var minValue = 25;
    return Calculate(minValue);
}
```

Bold: Indicates a new term, an important word, or words that you see onscreen. For instance, words in menus or dialog boxes appear in **bold**. Here is an example: "If you agree with the details and size of the installation, you can proceed to start the process by clicking on the **Install** button."

> **Tips or Important Notes**
> Appear like this.

Get in touch

Feedback from our readers is always welcome.

General feedback: If you have questions about any aspect of this book, email us at `customercare@packtpub.com` and mention the book title in the subject of your message.

Errata: Although we have taken every care to ensure the accuracy of our content, mistakes do happen. If you have found a mistake in this book, we would be grateful if you would report this to us. Please visit www.packtpub.com/support/errata and fill in the form.

Piracy: If you come across any illegal copies of our works in any form on the internet, we would be grateful if you would provide us with the location address or website name. Please contact us at copyright@packt.com with a link to the material.

If you are interested in becoming an author: If there is a topic that you have expertise in and you are interested in either writing or contributing to a book, please visit authors.packtpub.com.

Part 1: Visual Studio Overview

In this part, you will learn how to install and use Visual Studio from scratch and use the general tools provided by this IDE.

This part contains the following chapters:

- *Chapter 1, Getting Started with Visual Studio 2022*
- *Chapter 2, Configuring the IDE*
- *Chapter 3, Improvements in Visual Studio 2022*
- *Chapter 4, Creating Projects and Templates*
- *Chapter 5, Debugging and Compiling Your Projects*

1

Getting Started with Visual Studio 2022

Visual Studio (VS) is the most popular **integrated development environment (IDE)** for **.NET** developers. It's the perfect tool to design, develop, debug, and deploy all .NET applications and even other technologies.

In this chapter, you will learn about the history, historical versions, and installation process of VS, as well as the initial configurations to start working with this IDE. After learning about the VS flavors, you will be able to choose the option that best suits your needs.

By the end of this chapter, you will get a brief overview of VS's history and understand the main evolutionary changes across its different versions. You will also learn how to install and start using this amazing tool.

In this chapter, you will learn about the following main topics:

- A brief history of VS
- VS flavors
- Installing VS 2022
- VS for Mac

Technical requirements

We will begin the chapter by learning how to install VS 2022. To get VS to run on your machine, you will need the following requirements:

- Windows 10, version 1909 or higher

- Windows Server, 1016 or higher

- A 1.8 GHz or faster 64-bit processor; quad core or better is recommended

- 4 GB of RAM; 8 GB is recommended

- Hard disk free space – 25 GB (up to 40 GB depending on the components installed)

- Administrator rights

- Full internet access during the installation

> **Important Note**
>
> 32-bit and ARM operating systems are not supported; you will need either Windows 10 Enterprise LTSC edition, Windows 10 S, or Windows 10 Team Edition. To check all the requirements and technologies or systems not supported, go to `https://docs.microsoft.com/en-us/visualstudio/releases/2022/system-requirements`.

A brief history of VS

VS 2022 is version 13 of this application created by Microsoft. VS has been consolidated among developers for having a friendly user experience, good support with regular updates, and powerful tools for writing clean and scalable code. VS has support for many technologies and platforms. For many developers, VS is the ultimate tool for all project types.

To understand the evolution of this tool, we must examine its history and timeline.

VS 6.0 was released in 1997, and it was the first version of this tool. This version was created to work with Visual Basic 6.0. Then, in 2002, a new version was released, which included compatibility with **.NET** and **C#** (a new programming language at that time). Since then, it's been the favorite tool for .NET developers.

VS started as a premium application with a closed license, but since VS 2005, Microsoft began a new strategy with a freemium (free/premium) version, which is a basic/free public version that you can use for your personal projects, study, or midsize applications, and other versions at a cost for professional developers, large companies, or for those who want to use advanced tools.

Microsoft releases a new VS version every 2 or 3 years and provides updates for that version every 2 or 3 months, which means complete support.

A version of note is VS 2012 because the development team implemented a new look and feel and many improvements in the user experience, which are also present in the 2022 version. Some of the most important improvements in VS 2012 over previous versions were performance, the possibility to choose from light and dark themes, and new icons.

Now that you have a general idea of what VS is, let's examine each of the flavors available today.

VS flavors

Since version 2012, VS has had three flavors that cover all developers' preferences and needs. Just one of these three alternatives is completely free for the community.

In the following sections, you will see the differences between each version and learn about the features supported by each version, which are the following:

- Visual Studio Community
- Visual Studio Professional
- Visual Studio Enterprise

So, let's understand the main aspects of each version.

Visual Studio Community

Visual Studio Community is a free version that incorporates all the basic tools to create, build, debug, and deploy .NET applications and all the collaboration instruments integrated into VS.

Visual Studio Community has a limit of five users and is restricted to non-enterprise organizations.

The main tools in Visual Studio Community are the following:

- Basic debugging tools (tools for inspecting code during debugging)
- A performance and diagnostics hub (tools to analyze application performance and memory use)
- Refactoring tools (tools to clean and style code following best practices)
- Unit testing (a feature to navigate, run, and collect results from unit tests)
- Peek definition (a functionality to navigate to the definition of a method or function)
- VS Live Share (a tool for real-time collaboration development)

This version is suitable for students, independent developers, freelancers, and small companies. Even though this version includes all the main tools that you will use on a daily basis, in some scenarios associated with unit testing, memory, or inspection, these tools aren't enough.

Visual Studio Professional

Visual Studio Professional is a licensed version of VS offered by subscription; this version is recommended for enterprise applications and teams with more than five developers. VS Professional includes the same tools as Visual Studio Community but with some additions, such as CodeLens (a VS feature to find references, changes, and unit testing in code).

At the time of writing, the cost of VS professional subscription is $45 per month for an individual user.

The professional subscription includes $50 in Azure credit, training, support, and Azure DevOps (basic plan).

Visual Studio Enterprise

Visual Studio Enterprise is the top-level subscription version of VS (with Visual Studio Professional) that includes all of VS Community's features, Visual Professional's improvements, and some additional tools.

Some features to highlight are as follows:

- Live unit testing (a feature wherein unit testing is rerun every time a change is made)

- The Snapshot Debugger (a tool for saving snapshots during debugging when an error occurs)

- Performance analysis tools for mobile applications

- Architectural layer diagrams (to visualize the logical architecture of your app)

Enterprise subscription has a $250 fee per month, but it includes $150 in Azure credit, Power BI Pro, Azure DevOps with test plans, and all the features available for VS.

To see a comparison of the different flavors and prices, you can go to `https://visualstudio.microsoft.com/vs/pricing/`.

> **Important Note**
>
> For this book, we are going to use Visual Studio Community. Since this is a free version, you don't have to pay any subscription, and all the topics are covered with this version.

Installing VS 2022

The VS installation process has improved with each new version that has been released. Today, it is even possible to reuse the same installer to perform upgrades or workload modifications to create different types of projects.

In this section, you will learn how to do the following:

- Get the installer from the VS website

- Install VS

Let's see how to carry out the installation process in detail.

Getting the installer from the website

VS is easy to install, and you don't need an account or to take a lot of steps to get the installer. By typing `Visual Studio community Download` in any browser, you can easily find the link to download VS from the official page.

From the following link, you can directly get the installer:

`https://visualstudio.microsoft.com/downloads/`

On the webpage, go to the **Visual Studio** section, click on the drop-down control marked **Download Visual Studio,** and select the **Community** option (see *Figure 1.1*):

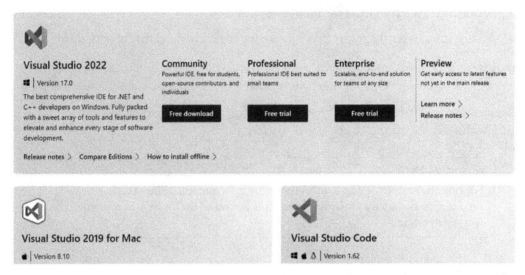

Figure 1.1 – Downloading VS

Installation process

Once you have downloaded the VS installer, you must run it to start the update process of the installer itself, as shown in *Figure 1.2*:

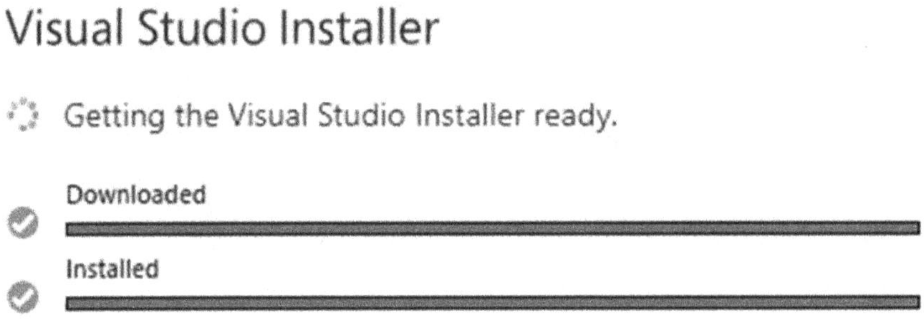

Visual Studio Installer

Getting the Visual Studio Installer ready.

Downloaded

Installed

Figure 1.2 – Updating the VS installer

VS is an IDE that is constantly updated by the development team, so the installer will always look for the latest available update to perform the corresponding installation.

When the installer update has been completed, the initial screen of the installer will be presented, which is composed of four main sections, with the **Workloads** section the one that is displayed by default.

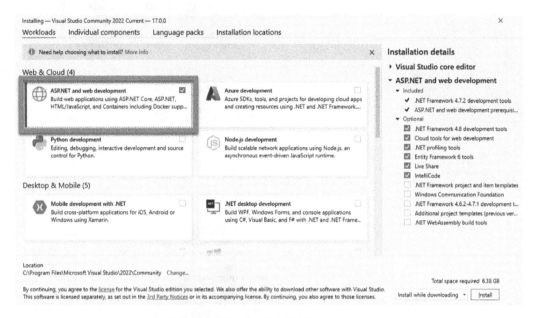

Figure 1.3 – The VS Workloads tab

It is possible to find workloads for different types of projects – for example, mobile projects, desktop-focused projects, and Python-focused projects, among others.

To install the templates and tools related to web development, you must select the workload called **ASP.NET and web development**, as shown in *Figure 1.3*.

Each workload includes a set of tools and components associated with the technology selected.

However, it is also possible to select these components individually, within the tab called **Individual components**, as shown in *Figure 1.4*:

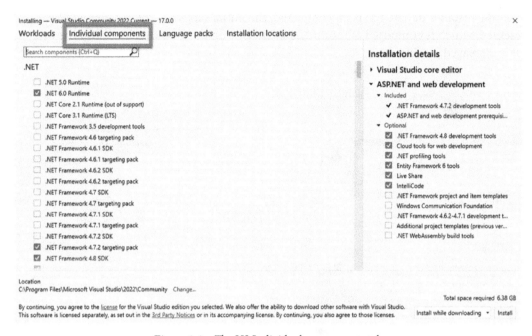

Figure 1.4 – The VS Individual components tab

These components are grouped into categories, such as .NET, clouds, databases, and servers, among others, and you only need to select a component to add it to the installation.

Within the tab called **Language packs**, you will be able to select the language or languages of the VS interface, as shown in *Figure 1.5*:

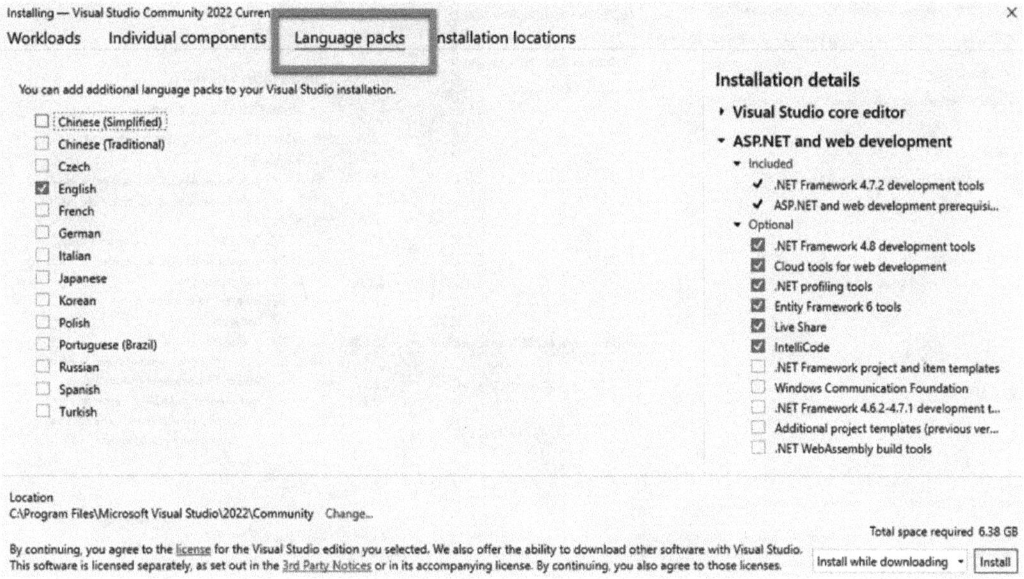

Figure 1.5 – The VS Language packs tab

This is very useful, since the default language usually corresponds to the language in which the VS installer has been downloaded. From here, you will be able to deselect the default language and select a different one or multiple languages to switch between in your development process.

In the last tab called **Installation locations**, you can configure the system paths of both the VS IDE and the download cache, as shown in *Figure 1.6*:

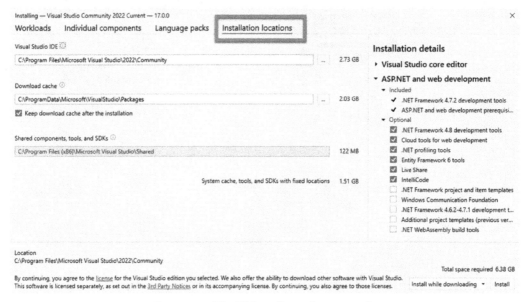

Figure 1.6 – The VS Installation locations tab

In this tab, you will be able to see how much space is needed in each of the paths to perform a correct installation.

Finally, there is a list of options to choose the installation method – whether you want to proceed to download all the components and install them at the same time, or you want to download all the necessary components first and install them later (*Figure 1.7*):

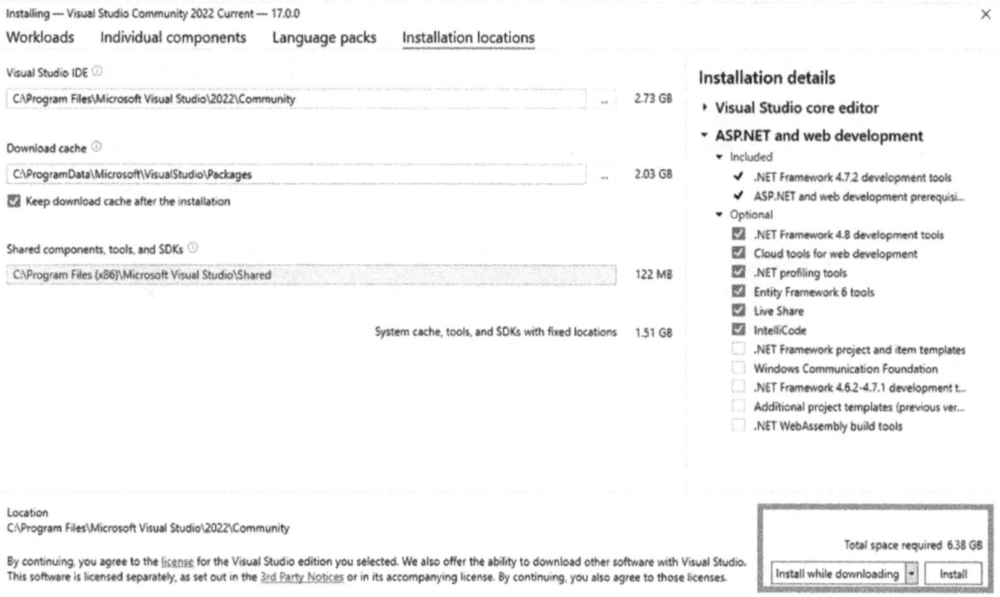

Figure 1.7 – The VS Install button

If you agree with the details and size of the installation, you can proceed to start the process by clicking on the **Install** button.

Once you have clicked the button to start the installation, a window will appear showing the details of the download and installation of the components, as shown in *Figure 1.8*:

Figure 1.8 – The VS installation in process

Once the installation process has finished, a window will appear, suggesting that you log in with a Microsoft account to get a license for the use of the tool. It is recommended that you log in at this time so that you do not lose access to the tool in the future, as shown in *Figure 1.9*:

Figure 1.9 – The VS login window

After you have obtained a license to use VS 2022, you will be shown the startup screen for creating, cloning, and opening projects, so you can check that the installation has been successful, as shown in *Figure 1.10*:

Figure 1.10 – The VS startup window

> **Important Note**
>
> You can have different versions of VS on the same machine; for compatibility with all technology, sometimes we need to keep old versions – for example, VS 2010 to work with Silverlight (an unsupported framework used to create web applications with C# and XAML, which was executed through a plugin in the browser).

After installing VS and getting the first screen, you can open VS using the **Continue without code** option.

You will now see the main screen in VS without having opened or created any project. For now, you only need to know that you can check the version and VS documentation using the **Help** menu (*Figure 1.11*):

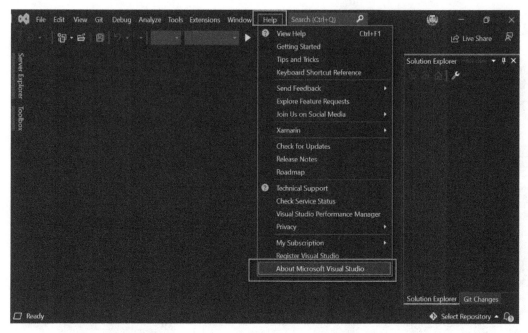

Figure 1.11 – The VS main screen and help menu

In *Figure 1.11*, the **About Microsoft Visual Studio** option is highlighted; by clicking on it, you will know the version being used and whether there is any update to install.

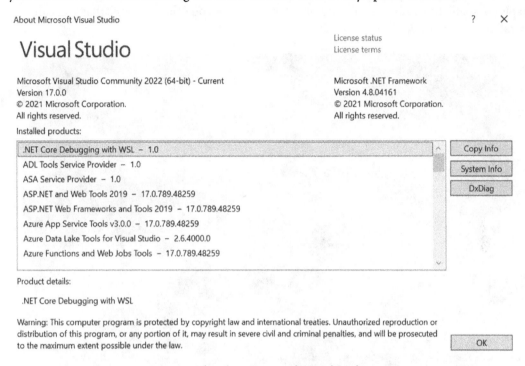

Figure 1.12 – The About Microsoft Visual Studio screen

Finally, if you need to repair, modify, or uninstall VS, you can use the VS installer that you downloaded earlier, or search for the term `Visual Studio Installer` in Windows Explorer at any time. Also, it will find updates and show you other versions that you can install or try (*Figure 1.13*):

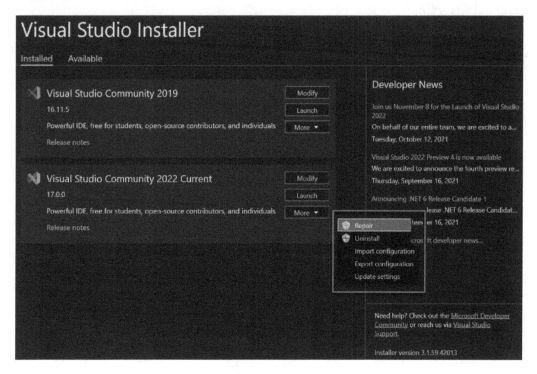

Figure 1.13 – The VS Installer options

Now that we know the VS 2022 installation process, let's explore VS for Mac users.

VS for Mac

VS has a new version for Mac users. Using this version, developers can have a similar experience in MacOS as Windows users.

Due to incompatibility issues, which have to do with the operating system, .NET Framework is not supported in VS for Mac, only VS for Windows.

Figure 1.14 – VS for Mac – the startup screen

In *Figure 1.14*, you can see how VS for Mac looks when it is opened; it looks pretty much like the Windows version. The user can choose a recent project, open a project, or create a new one.

> **Important Note**
> Even though VS for Mac was made using technology for macOS operating system, the development team is working to maintain the same experience as VS for Windows.

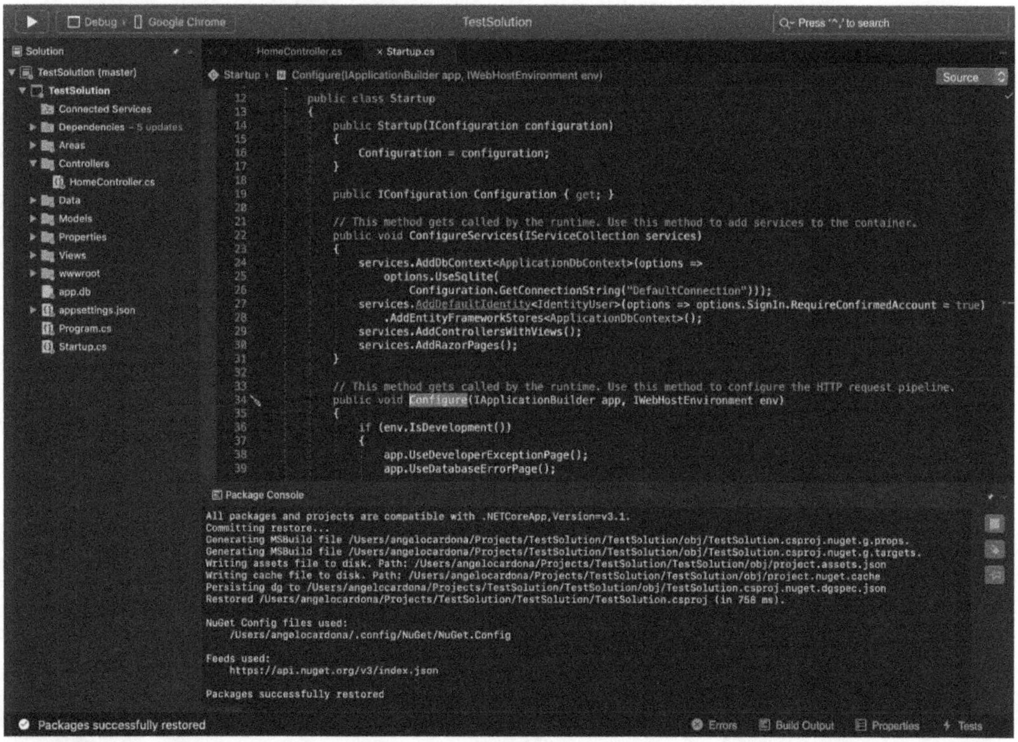

Figure 1.15 – VS for Mac – the main screen

After opening a project, you can set up your toolbar, but by default, VS for Mac has the **Solution Explorer**, **Properties**, and **Errors** sections. VS for Mac is an excellent option for developers that need debugging and performance tools as well as for .NET MAUI (technology to create native apps using C# and XAML or using Blazor) developers who want to create apps for Android, iOS, and MacOS.

If you want to try VS for Mac, you can get the installer at the following link:

`https://visualstudio.microsoft.com/vs/mac/`

Summary

In this chapter, you have learned what VS is, the different versions of the IDE available that you can choose to suit your project needs, the installation processes of VS 2022 for Windows, and the general concept of VS for Mac.

In *Chapter 2, Configuring the IDE*, you will learn how to customize the IDE to suit your needs as a web developer.

2
Configuring the IDE

In general, it is common for software developers to have different tastes in writing code. This also applies in the configuration part of the tool you use to develop, such as whether you want to change the general color of the IDE or prefer to change the default font to one that suits you better.

In this chapter, you will learn about the different configuration options in Visual Studio, which can certainly help you improve your productivity by having the exact elements you need, both in terms of colors and the location of windows that you frequently use.

By the end of this chapter, you will be able to modify the IDE color scheme, configure dockable windows, and customize fonts.

You will learn about the following topics:

- Synchronizing accounts and settings
- Configuring the color scheme
- Customizing fonts
- Customizing the menu and toolbars
- Customizing panels

Technical requirements

For this chapter, you will need to have Visual Studio 2022 already installed on your machine. In *Chapter 1*, *Getting Started with Visual Studio 2022*, you can check the requirements and the installation process.

Synchronizing accounts and settings

A great feature of Visual Studio is that it allows you to synchronize the configurations you make, and also allows you to comfortably work on different computers.

This is possible thanks to a Microsoft account, which is required for the use of Visual Studio. This account is requested when you start Visual Studio for the first time, or you can enter the account or modify it for use in Visual Studio at any time by clicking on the **Sign in** option, located at the top right of the IDE, as shown in *Figure 2.1*:

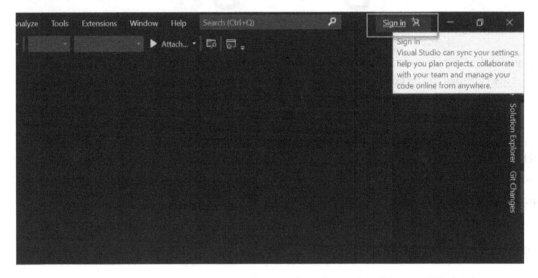

Figure 2.1 – Visual Studio – Sign in (at the top right of the main screen)

The main configurations that are synchronized through this process are as follows:

- User-defined window layout configurations
- Themes and menu settings
- Fonts and colors
- Keyboard shortcuts
- Text editor settings

If you do not want to apply the synchronization of your configuration on a particular computer, it is possible to do so by going to the **Tools | Options | Environment | Accounts** menu. From here, you will be able to deselect the **Synchronize Visual Studio settings across devices** option, as shown in *Figure 2.2*:

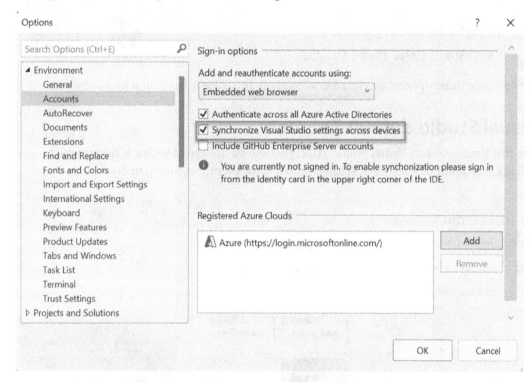

Figure 2.2 – Disable the synchronization of preferences between devices

> **Important Note**
> Disabling this feature will not affect the synchronization process of other versions or editions of Visual Studio that may be installed on the same computer.

Now that we have learned about synchronizing settings between devices, let's look at how to customize the main settings.

Configuring the color scheme

There are different ways to customize Visual Studio 2022, one of the most important of which is to adjust colors.

Whether you're in a well-lit space or coming from another programming environment, you'll probably want to adjust colors according to your personal preferences.

To customize the color scheme in Visual Studio 2022 to our liking, there are two things we can use:

- Visual Studio default Themes
- Visual Studio Color Theme Designer

Let's analyze these options and learn how to apply the color scheme that best suits you.

Visual Studio default Themes

The first time you start Visual Studio 2022, you will be presented with a window, which will ask you about development settings, and a color theme, as shown in *Figure 2.3*:

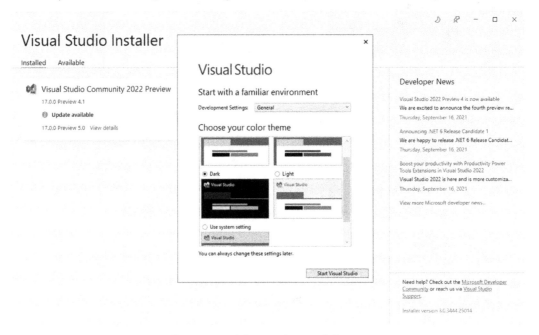

Figure 2.3 – Selecting the initial theme

These color themes are as follows:

- **Blue**
- **Blue (Extra Contrast)**
- **Dark**

- **Light**

- **Use system setting**

Each of these options has a set of pre-established colors that you will be able to preview using the same window.

Once you select the theme of your choice, it will be applied and saved in the configuration that is associated with the Microsoft account with which you have requested the license for the use of the IDE.

> **Important Note**
>
> The **Dark** theme helps reduce eye strain in low-light conditions. This is a perfect option if you need to work for many hours per day in an office or places with limited light. *All the figures and screenshots will be in dark mode in this book.*

In case you want to change the theme you selected at the beginning, you can do so by going to the main Visual Studio window and clicking on the link that says **Continue without code**:

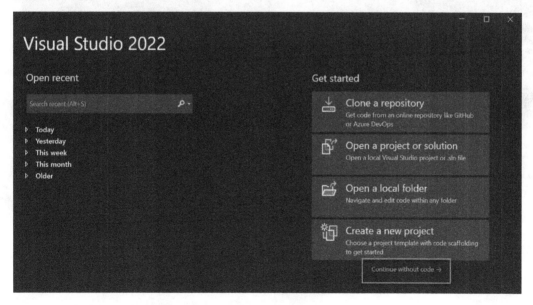

Figure 2.4 – Accessing Visual Studio without code

Then, from the **Tools** dropdown menu, go to the **Themes** section; you will find all the default themes and those that you have previously installed. You only need to select one to apply the selected theme, as shown in *Figure 2.5*:

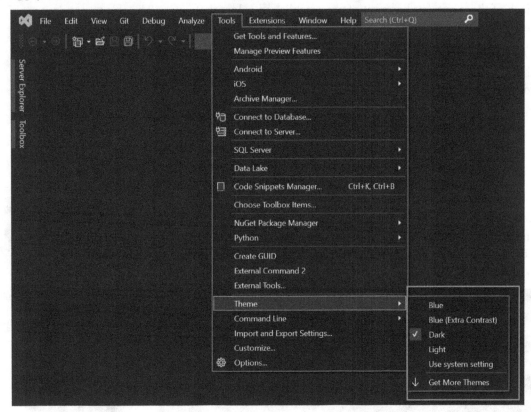

Figure 2.5 – Changing the Visual Studio theme

There is another way to change the theme. You can go to **Tools | Options | Environment | General** and choose **Color Theme**:

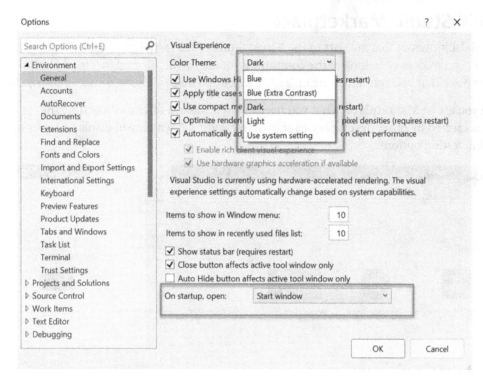

Figure 2.6 – Changing the color theme in the Options section

In this menu, you can also change how Visual Studio starts by selecting the different options in the section marked **On startup, open**. You can choose between the following options:

- **Start window**: The default window where you can select a recent project or create a new one (see *Figure 2.4*).

- **Most recent solution**: Visual Studio will start with the last project or solution that was opened.

- **Empty environment**: Open the main windows in Visual Studio without selecting a project or solution.

Let's now review a fabulous option to download additional themes.

Visual Studio Marketplace

If the default themes that are part of the Visual Studio 2022 installation are not enough for you, you may want to look at the community-made themes hosted in Visual Studio Marketplace that you can find at `https://marketplace.visualstudio.com/`.

Once you are in **Marketplace**, what you need to do is make a filter to show only the themes created by other developers, as shown in *Figure 2.7*, which will display a series of very interesting options:

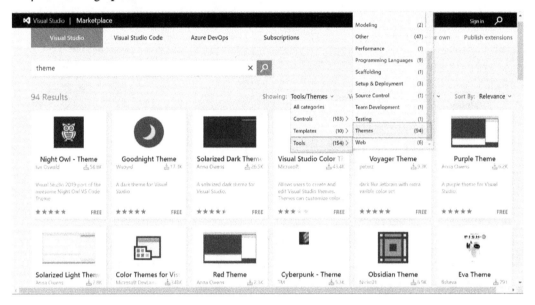

Figure 2.7 – Searching for themes in Visual Studio Marketplace

If you want to know more about a specific theme, you need to click on the element, which will show you information about the theme, and almost always, an image of what the theme will look like once it is applied to Visual Studio will appear.

If you want to install the theme, just click on the **Download** button, which will download the corresponding installer, and when you run it, it will start the installation process of the new theme.

Once the theme has been installed, just follow the same steps as in the previous section to switch to the new theme.

> **Important Note**
> Within the same Marketplace, there is a plugin called Visual Studio Theme Designer, which allows you to easily create your own themes.

Now that we have seen how to select a theme that fits our visual preferences, let's see how we can customize the font styles as well.

Customizing fonts

There are two places where you will probably be interested in changing the type of font you use:

- In the general environment

- In the source code editor

Let's analyze each of the options as follows.

Changing fonts in the IDE

To change typography at the IDE level, follow these steps:

1. Go to the **Tools | Options** menu.

2. In the configuration window, go to the **Environment | Fonts and Colors** section.

3. In this section, select the **Environment** option in the **Show settings for** drop-down list.

   This will allow you to change options such as font and size:

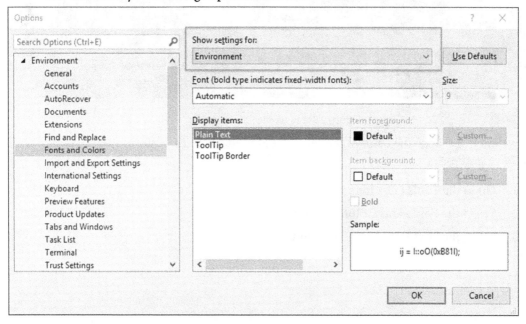

Figure 2.8 – Changing the environment font settings

In *Figure 2.8*, we can use the option **Font** to change the **Automatic** or **Default** font in Visual Studio and see how it looks using the **Sample** section.

Now let's review reference highlighting to see how to customize the way Visual Studio highlights text in the code.

Reference highlighting

Another common customization is to change the highlighting of references. This refers to highlighting the different occurrences of a selected element, such as a variable or keyword. This can be best seen in *Figure 2.9*, in which, when positioned over the int keyword, all references to it are highlighted:

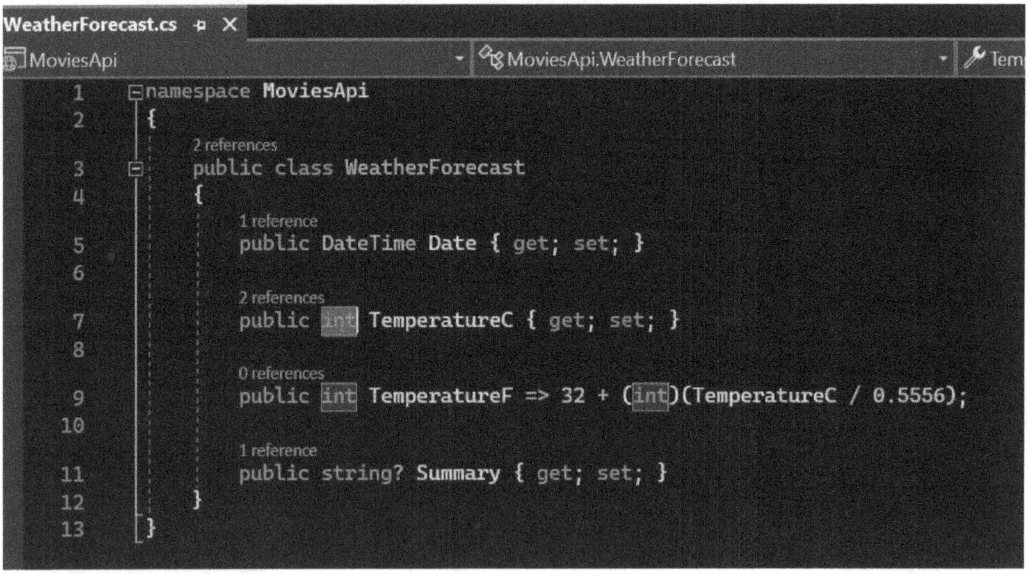

Figure 2.9 – Reference highlighting

If we want to change the color of the references found, we must perform the following steps:

1. Go to the **Tools | Options** menu.

2. Select the **Fonts and Colors** section.

3. Select the **Text Editor** configuration.

4. In the **Display items** section, select the configuration called **Highlighted Reference** and change the corresponding colors, as shown in *Figure 2.10*:

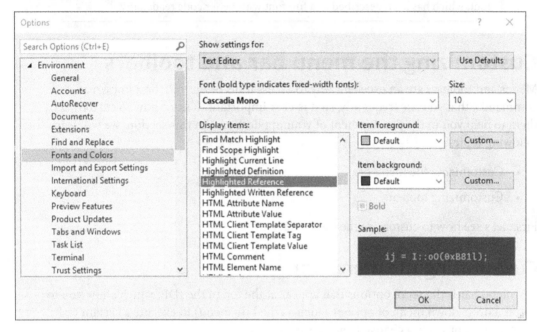

Figure 2.10 – Changing the highlighted references options

You can choose the text color in **Item foreground** and the backgound color in **Item background**, and finally see a preview of the color selected using the **Sample** section.

Changing fonts in the code editor

To make changes to the typography used in the text editor, follow the same steps shown just before *Figure 2.10*.

As part of the configuration options, you can change the font used, the font color, the font size, and the background, among other attributes.

One of the advantages of this configuration option is that it allows you to be as specific with the configuration as you need. This means that you can also alter the typography for things such as selected text, line numbers, bookmarks, and code snippets, among many other settings.

Now that you know how to change the typography to suit your needs, let's see how to adjust the panels within the IDE so that your workflow will be faster according to your projects.

> **Important Note**
> In Visual Studio 2022, the Cascadia font has been introduced as the default font, which has been described as a fun font with better code readability.

Customizing the menu bar and toolbars

Menus and toolbars are an excellent way to access tools or options, best known as commands, that you use frequently, so it is very important to learn how to customize them to help you in the development of your applications. In this section, we will learn the following topics:

- Customizing the menu bar
- Customizing toolbars

First, let's see how to customize the menu bar.

Customizing the menu bar

The menu bar is the set of options that appear at the top of the IDE, which allow you to access a drop-down menu of options (such as the **File** menu) to execute a certain task, display tools, or apply a change to a project:

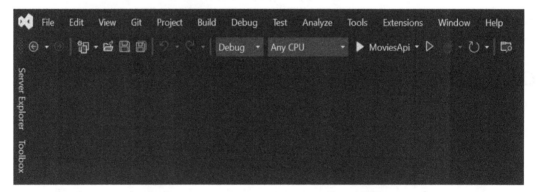

Figure 2.11 – The default menu bar in Visual Studio 2022

If you want to change the tools that are part of the initial configuration, either to add options in a specific menu or to create your own menus, perform the following steps:

1. Go to **Tools | Customize**.
2. Go to the **Commands** section.

3. In this section, you should work with the **Menu bar** option, which will allow you to modify a main menu bar and a secondary menu bar, which you can differentiate with pipe symbols (|) in the dropdown, as seen in *Figure 2.12*:

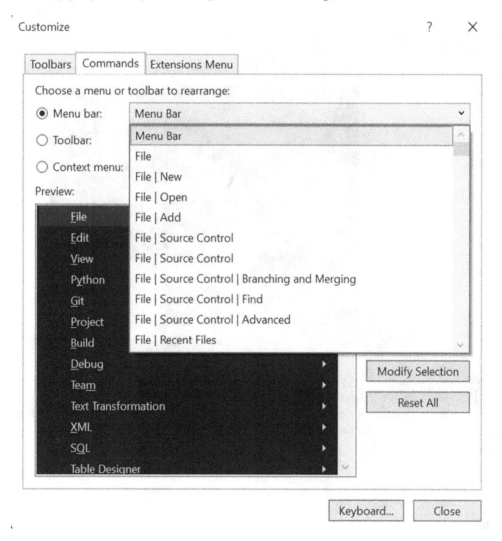

Figure 2.12 – The Menu bar customization window

4. Once you have selected the menu you wish to modify, a preview will appear, showing you how the menu currently looks. From here, you can add new commands to the menu bar by clicking the **Add Command** button.

This will open a new window that will show you each of the commands grouped by category, which you can select to add to the selected menu, as shown in *Figure 2.13*:

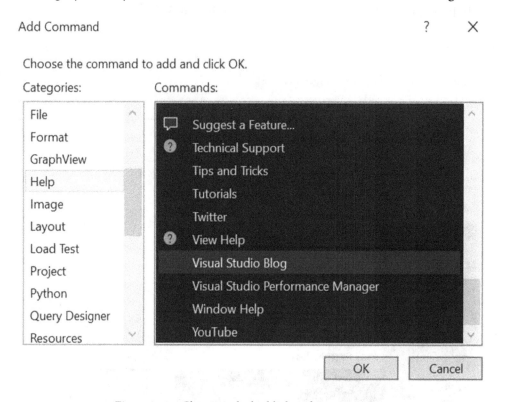

Figure 2.13 – Changing the highlight references options

If you want to remove the option added to the menu, just select it and click on the **Delete** button, which is highlighted by a rectangle in *Figure 2.14*:

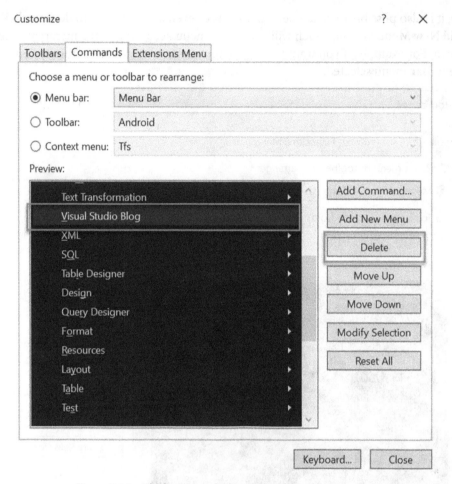

Figure 2.14 – The button to delete an item from the menu bar

You also have the **Move Up** and **Move Down** buttons, which allow you to move an option up or down on the menu and rearrange the menu items.

Important Note

It is possible to delete toolbars created by the user but not those that are part of the default configuration.

Finally, it is also possible to create new submenus or even a new menu. To do this, click on the **Add New Menu** button, which will add a new menu according to the hierarchical level you are in. For example, if you want to add a main menu, you must select this option with the **Menu Bar** menu selected:

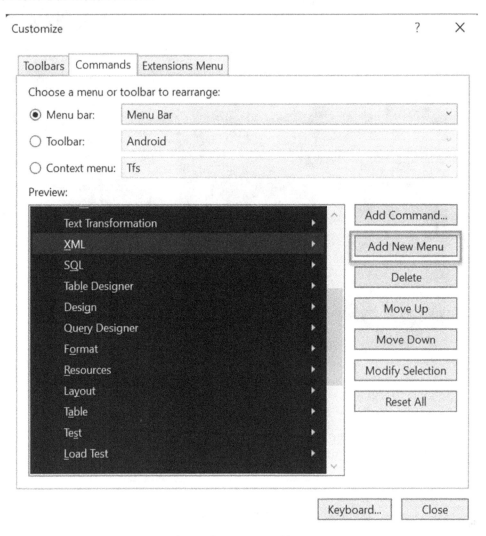

Figure 2.15 – The configuration to add a new main menu

Conversely, if you want to add a submenu – for example, in the **Edit** menu – you will have to carry out this action having selected the **Edit** menu:

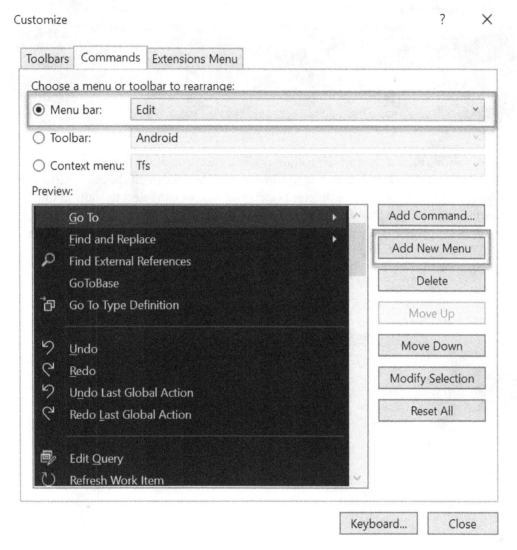

Figure 2.16 – The configuration to add a submenu in the Edit menu

The customization of toolbars is done in the same window; let's see how to do it next.

Customizing toolbars

The toolbar is the set of commands that you can access directly without the need to open a menu first, as shown in *Figure 2.17*:

Figure 2.17 – The Visual Studio 2022 toolbar section

It is also possible to configure which commands will appear in different tool groups by selecting **Tools | Customize**. In this window, we will see by default the tab called **Toolbars**, which will show us the different categories that we can select to show in the interface. By default, the **Standard** option is selected for API and web projects. Depending on the projects, there are other toolbars added by default, but we can add other toolbars manually, simply by selecting them with a tick, as shown in *Figure 2.18*:

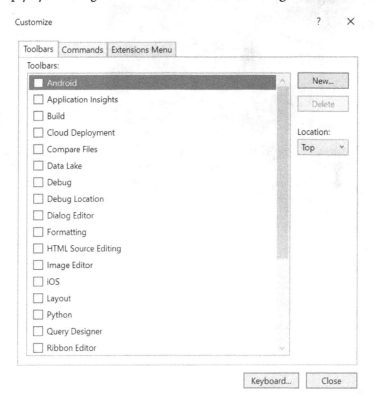

Figure 2.18 – The customization window for adding toolbars to the toolbar section

You can add new commands to a toolbar by going to the **Commands** tab. In this tab, we will select the toolbar we want to modify and carry out the same steps described in the *Customizing the menu bar* section.

If you want to quickly add different toolbars to the IDE, *Figure 2.19* shows how to do it easily by accessing the **View** menu and **Toolbars**, where you can select and deselect the toolbars you are interested in:

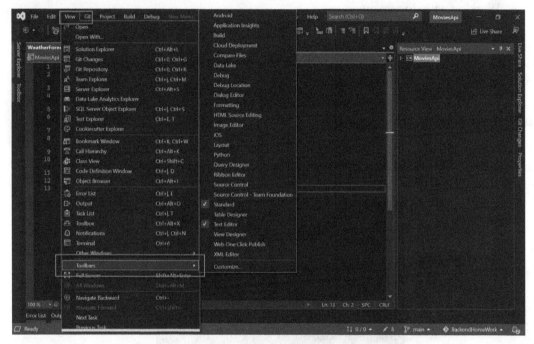

Figure 2.19 – Access to the toolbars from the View menu

You now know how to add and remove toolbars in Visual Studio to customize the set tools depending on your preferences or technologies you use. Next, let's review how to customize panels in Visual Studio.

Customizing panels

Panels in Visual Studio are a way to access specific tools according to the type of project you are working on.

These panels are composed of tools and document editors, some of which are used most of the time, such as the solution explorer (to see the structure of your projects), the toolbox (which shows you controls to drag and drop according to the current project), the properties panel (to modify the properties of the selected element), and the code editor.

It is important that you know how to work with these panels so that you can configure the set of tools and editors that best suits your projects. That is why in this section, we will look at the following topics:

- Adding tools to panels
- Panel accommodation
- Working with documents
- Managing window layouts

Let's learn how to configure the IDE to show you the tools that you will need for your project.

Adding tools to panels

Visual Studio 2022 has many tools or windows that you can show or hide according to your needs. This list of tools can be found in the **View** menu, as demonstrated in *Figure 2.20*:

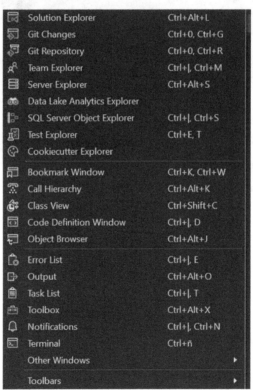

Figure 2.20 – The list of available tools

Once you have opened this menu, you will find the tools that are usually the most used listed immediately. These tools include tools such as the server explorer, the class viewer, the error listing, the output window, and the terminal, among other powerful tools.

There is another set of tools that are not as widely used but may help you at some point. These are found in the **Other Windows** section.

From here, you can access tools such as containers, C# interactive, Data Sources, and the Package Manager Console, among others.

To add any of these tools to one of the panels, simply select one of them, and it will automatically be added to your current environment in a strategic panel. For example, if you add the **Server Explorer** tool, it will be added to the left panel. On the other hand, if you add the **Output** tool, it will be added to the bottom pane.

Panel accommodation

A great advantage of Visual Studio is that you can arrange the tool panels wherever you prefer. To achieve the best results, it is convenient that you know the structure of a panel.

Each panel is composed of five sections where you can place tools. These sections are located on each side of the panel, plus one in the center, as shown in *Figure 2.21*:

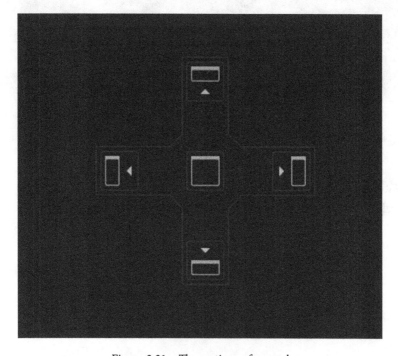

Figure 2.21 – The sections of a panel

To place a tool in a panel, simply position the cursor at the top of the tool and slide it to the panel you want. The IDE itself will show you the possible locations where you can place the tool, making the process simple and easy; you can also pull the panel out of the main windows and use it on its own. In *Figure 2.22*, you can see what the process of docking a window looks like:

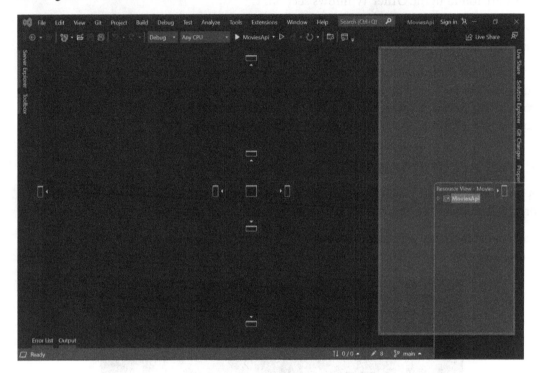

Figure 2.22 – Docking a window in a panel

Now, it's time to review how we can work with documents.

Working with documents

There are special options that we can apply when working with document editors, such as the code editor. If we want to see these options, we just need to right-click on the tab of the open document, as shown in *Figure 2.23*:

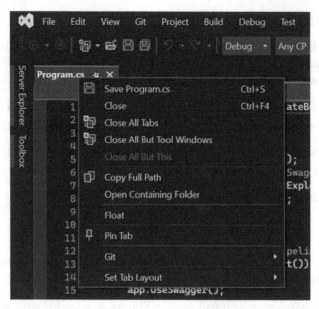

Figure 2.23 – Changing the environment font settings

These options are quite intuitive – for example, the **Float** option will allow us to turn the editor into a floating window, which we can drag to a second monitor. The **Pin Tab** option will allow us to set the tab at the beginning of the open windows, and the **Set Tab Layout** option will allow us to move the set of tabs to the left, the top, or the right.

It is important to highlight that if we have more documents open, we will have additional options available. With this option, we can create groups of documents to distribute the space and use it for performing tasks to increase productivity, such as comparing two documents. *Figure 2.24* demonstrates this visually:

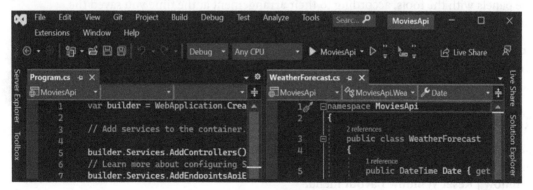

Figure 2.24 – Document groups

Now you know how to manage your documents in the editor and organize them. Next, let's learn about managing window layouts.

Managing window layouts

It is very likely that in your life as a developer, you will come across projects of all kinds. You may have to manage things related to databases in a project, so you will probably want to keep the **Server Explorer** tool open.

It may be that in another project you are working on at the same time, you don't have to touch anything related to databases and you want to keep the toolbox open.

Is there any way to have tools located in the panels you prefer, without cluttering the IDE with tools? The answer is yes, thanks to the use of window layouts.

Let's see the main tasks related to them.

Saving window layout

Once you have configured your panels with the tools you require for a certain project, you need to go to the **Window | Save Window Layout** menu.

This will open a new window, which will prompt you for a name for your workspace configuration. Once you have entered the name, the configuration will be saved automatically.

To verify that the change has been made, you can go to the **Window | Manage Window Layouts** menu, which will show you a window with all your previously saved layouts or workspaces.

Applying window layout

Once you have saved at least one layout, you can apply it so that your workspace loads the panels with the tools, according to their arrangement at the time you saved the window layout.

To apply a layout, you must go to the **Window | Apply Window Layout** menu, which will show you all the previously saved layouts. Select the one you want to apply, and you will have your workspace loaded.

Resetting a window layout

Finally, you may want to reverse all changes made to a workspace so that you have the initial Visual Studio configuration. Fortunately, this option is available by going to the **Window | Reset Window Layout** menu.

When we press this button, we will be asked whether we agree to return to the initial configuration. We just need to click on **Yes** to have the environment with the default configuration.

> **Tip – a Help Shortcut**
> Use *Ctrl + F1* to open the Visual Studio documentation to read guides and news.

Summary

In this chapter, you have learned how to customize Visual Studio, choosing a theme, fonts, colors, and panels. Depending on your needs, you can not only set up your Visual Studio preferences but also try default configuration settings first, and adjust Visual Studio according to your work style.

In *Chapter 3, Improvements in Visual Studio 2022*, you will learn about the new improvements added in this release and see a comparison with Visual Studio 2019.

3
Improvements in Visual Studio 2022

Every new Visual Studio version brings new features to improve user experience and performance and provides more functionalities and tools for developers.

We are just starting with Visual Studio 2022. In this book, you will learn some features with medium and high complexity, and in later chapters, we will study the changes in Visual Studio 2022 in more detail to get a deeper understanding. But in this chapter, we will review the most important features available in Visual Studio 2022 that you can use now.

By the end of this chapter, you will recognize the main differences between Visual Studio 2019 and Visual Studio 2022 and the most relevant improvements in this new version.

We will review the following improvements in Visual Studio 2022:

- 64-bit architecture
- New icons and styles
- .NET 6 support
- Hot Reload
- Other improvements

Technical requirements

For this chapter, you will need to have Visual Studio 2022 already installed on your machine. In *Chapter 1*, *Getting Started with Visual Studio 2022*, you can check the requirements and the installation process.

64-bit architecture

A simple but important feature in Visual Studio 2022 is the new architecture in 64 bit. This is a change that we cannot see, but internally, it takes advantage of a 64-bit CPU to improve the performance and reduce delays in the execution of multiple tasks.

Visual Studio 2022 only supports a 64-bit system; this is something common in current laptops and PCs.

Using **Task Manager** in Windows, you will be able to notice the difference when Visual Studio 2019 and 2022 are running at the same time:

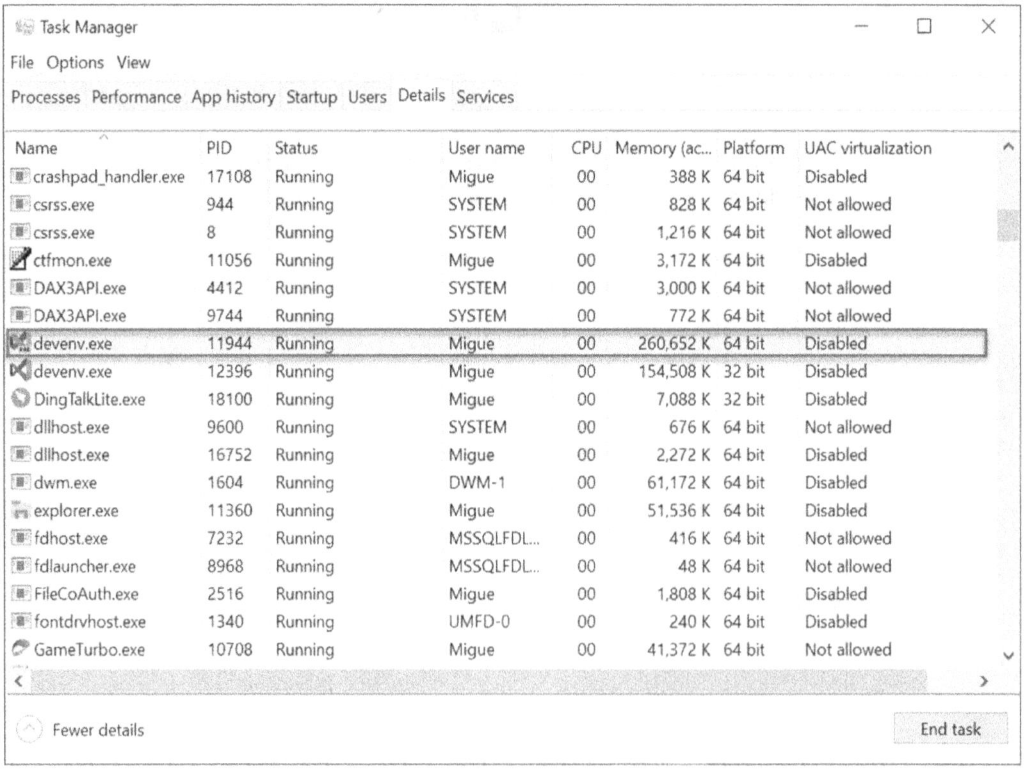

Figure 3.1 – Visual Studio 2022 on the 64-bit platform and 2019 on the 32-bit platform

With a 32-bit architecture, there was an access limitation of 4 GB of memory. Now, thanks to a 64-bit architecture, it is possible to access a larger amount of memory, reducing time limits and avoiding IDE freezes or crashes.

All in all, the Visual Studio development team improved the performance for many scenarios in version 2022. You will notice the difference when working on large projects, and performance will be better in future versions.

64 bit is a good improvement for performance, but this feature doesn't improve user interaction while coding. In the following section, we will see how the icons and style were improved to have a better user experience.

New icons and styles

New icons and styles were added in Visual Studio 2022. Although this is a simple feature, it helps us to navigate easily in an application, using visual elements and identifying actions and tools properly. For instance, in *Figure 3.2*, you can see the broom icon (row one, column five) has better contrast, with a new vibrant yellow color (specifically for a dark theme) and a modern design. It also implies that a code cleanup will be executed. The broom icon is used to execute code cleanup to fix code formatting:

Figure 3.2 – Icons in Visual Studio 2022 versus 2019

Important Note

An interesting fact about the new icons in Visual Studio 2022 is that the Visual Studio development team worked with the developer community to fulfill three purposes – consistency, readability, and familiarity. This resulted in a series of icons with the same meaning but with consistent colors, sharp contrast, and a recognizable silhouette.

The contrast between letters, icons, and background was improved to make it more pleasant and less tiring for the eyes. In *Figure 3.3*, you can see an example:

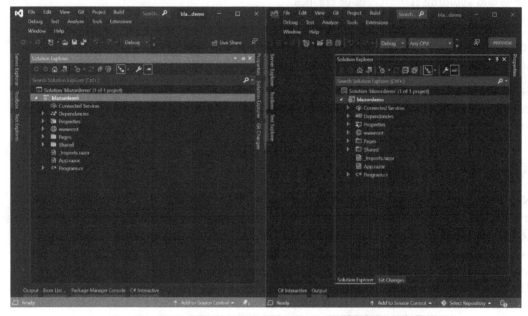

Figure 3.3 – An example of a dark theme in Visual Studio 2019 (left) versus 2022 (right)

> **Important Note**
>
> While the use of dark themes is becoming more common among developers in general, it is also becoming more prevalent in applications across the industry.

Figure 3.4 is another example of using the blue theme. Although the colors are very similar, we can see some differences and how the screen looks better with the new icons:

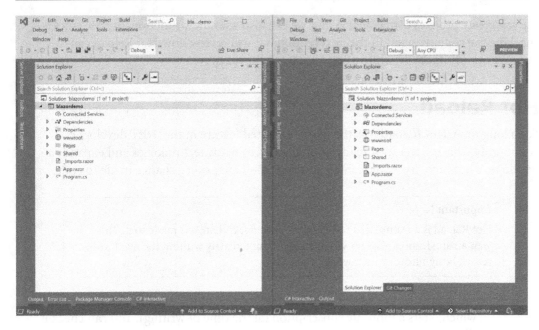

Figure 3.4 – An example of a blue theme in Visual Studio 2019 (left) versus 2022 (right)

There are also some changes in the other themes, but dark and blue were the most improved.

With these new icons and styles, working with Visual Studio becomes more user-friendly. In the next section, we will see how Visual Studio supports the new version of .NET, .NET 6.

.NET 6 support

.NET is a free, open source framework used to create web, desktop, mobile, and other kinds of applications, using C#, F#, and Visual Basic (with C# being the most popular one).

A new version of .NET has been released every year. .NET 6 is a new long-term support version, which offers great new improvements, such as minimal APIs and C# 10 compatibility. For more information about the improvements, you can visit https://docs.microsoft.com/en-us/dotnet/core/whats-new/dotnet-6.

Visual Studio 2022 is ready to create, compile, and publish projects using this new version.

In *Chapter 4*, *Creating Projects and Templates*, we will analyze the projects and templates provided by Visual Studio and look at the option to choose .NET 6 for our applications.

We can create projects using .NET 6, but we can also use a new functionality for quickly checking the code changes. In the next section, we will review how Hot Reload can improve our productivity while using Visual Studio 2022.

Hot Reload

For a long time, **Hot Reload** was the main requested feature in the .NET developers' community. This is a feature already implemented in many technologies and expands a developer's productivity by refreshing an application after every change made to code.

> **Important Note**
>
> Hot Reload is a feature that recompiles code after a change is made to it. This way, an application displays visual changes immediately without the need to restart it, significantly increasing productivity.

Even external companies are working on this feature; one of the most popular is LiveSharp, which offers a monthly subscription (at the time of writing). You can check it out at the following link:

```
https://www.livesharp.net/
```

Visual Studio 2022 includes this amazing feature for many kinds of projects, including ones involving ASP.NET and Blazor (by creating web apps using WebAssembly and Razor components):

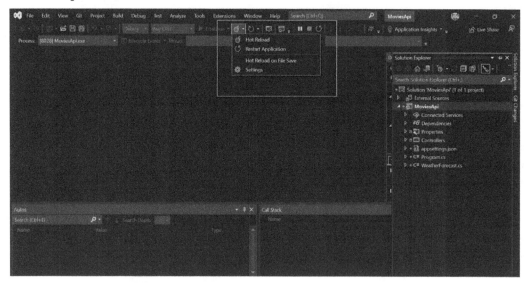

Figure 3.5 – The Hot Reload button in Visual Studio 2022

> **Important Note**
>
> The team behind Hot Reload decided not to support the new Hot Reload feature in *Xamarin.Forms* based projects for Android and iOS scenarios, so it is recommended to perform a migration to .NET MAUI to use this feature.

In later chapters, we will use this utility in some demos and learn why this tool is very important for us. In the next section, we will share other improvements that can help us to increase our productivity, especially in larger projects.

Other improvements

Visual Studio 2022 has other improvements, one of which is performance. Visual Studio 2022 runs on 64 bit, but in addition, features such as the search file tool were improved to help find elements quickly in a project with a large number of files. In the following diagram, you will see a search example:

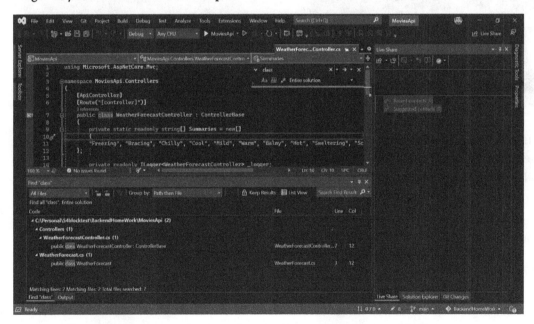

Figure 3.6 – An example of searching for the word "class" in Visual Studio 2022

Also, you have a new type of breakpoint (*a breakpoint is a technique to stop application execution in a specific line of the code*). With this feature, we can add a breakpoint that depends on another created before. We will see more information about this feature in *Chapter 5, Debugging and Compiling Your Projects*. In *Figure 3.7*, you can see the new **Insert Dependent Breakpoint** option:

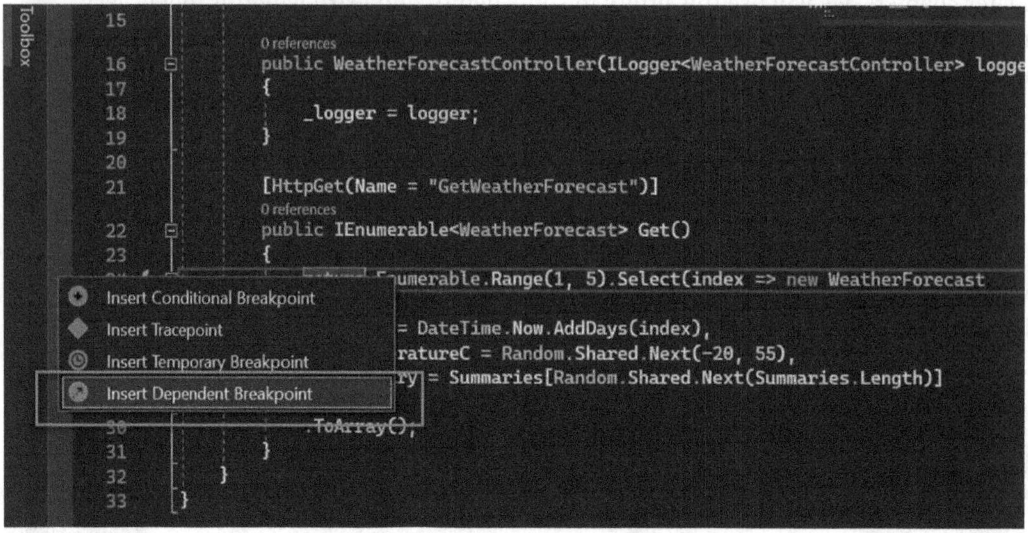

Figure 3.7 – The dependent breakpoint in Visual Studio

This is a new type of breakpoint that is very useful for debugging large applications with high complexity. It will interrupt the execution of your application when a previous breakpoint meets a condition.

Regarding Git integration, we have a new experimental feature that helps us to work with multiple repositories in the same solution. Thus, you can perform changes in different projects and then commit the changes (save the changes in Git) without opening multiple instances of Visual Studio. In *Figure 3.8*, the new option to enable multiple repositories is shown:

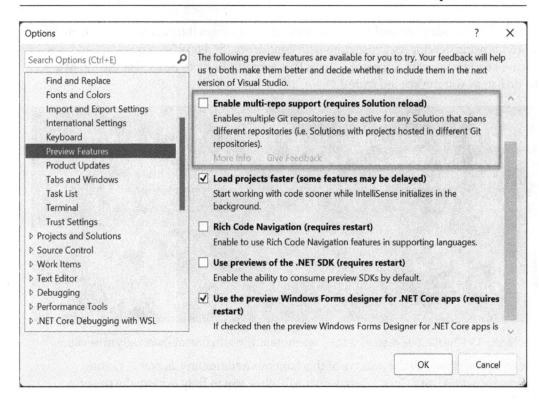

Figure 3.8 – In Tools | Options | Preview Features, there is a new option to enable multi-repo support

We will analyze this feature deeper in *Chapter 11*, *Implementing Git Integration*, where we will use a repository in GitHub to try functionalities related to the integration with Git and GitHub.

Finally, IntelliCode is undoubtedly one of the major changes introduced in Visual Studio 2022, allowing developers to program with confidence, find problems quickly, and have focused code reviews by suggesting items that are most likely to be used at the top of a to-do list, as you can see in *Figure 3.9*:

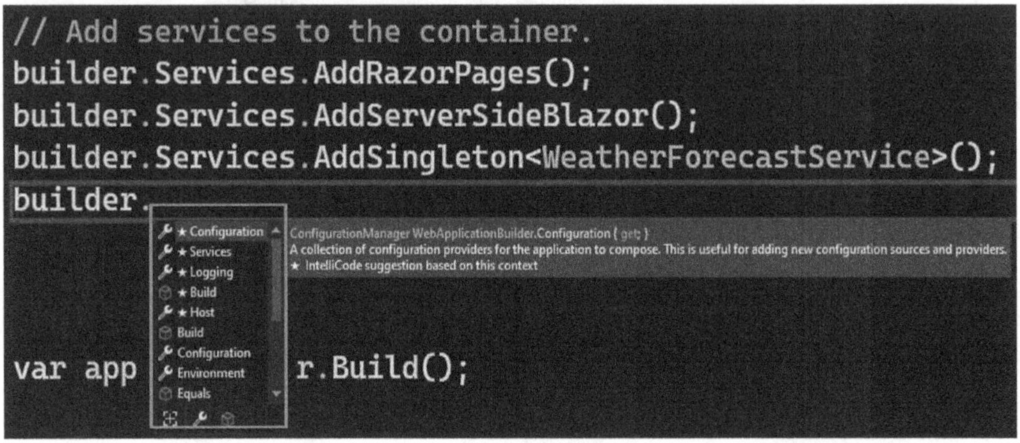

Figure 3.9 – IntelliCode suggesting the properties and methods that are most likely to be utilized

We will get to know all the features of this fabulous addition in *Chapter 7, Coding Efficiently with AI and Code Views*, which will allow you to fully explore the use of IntelliCode.

Summary

In this chapter, we provided an overview of the main improvements in Visual Studio and some differences between Visual Studio 2022 and Visual Studio 2019. Throughout this book, we will analyze these features further and use them in some demos.

In *Chapter 4, Creating Projects and Templates*, you will create your first project and analyze the templates provided by Visual Studio for different programming languages and technologies.

4
Creating Projects and Templates

A template within Visual Studio is a set of files, references, project properties, and compilation options for working with a particular technology. Templates provide us with basic code to work with and acts as a guide that we can follow and complete by including our business logic and requirements. Different templates are installed according to the workloads that have been selected during Visual Studio installation, although there are templates that will be installed by default, such as class libraries. Depending on the project or technology we want to use, we will find different template options to choose from. Selecting the right template that fits your needs is the best action that you can take to evade technical debt and future issues in your architecture.

In this chapter, we will analyze the main web development templates provided by Visual Studio 2022 and the options available to customize these templates. Also, during this chapter, we will understand how to pick the best template for our projects, considering the scope, requirements, and expertise of the team.

We will review the following topics in this chapter:

- Selecting and searching for templates
- Templates for .NET Core
- Templates for **application programming interfaces (APIs)**

- Templates for .NET Framework
- Templates for **single-page application** (**SPA**) projects

Let's see what these templates are all about and how to work with them.

Technical requirements

To follow along with this chapter, you must have previously installed Visual Studio 2022 with the web development workload, as shown in *Chapter 1*, *Getting Started with Visual Studio 2022.*

You can check out the changes made to the project at the following link: `https://github.com/PacktPublishing/Hands-On-Visual-Studio-2022/tree/main/Chapter04`

Selecting and searching for templates

As mentioned in the introduction of this chapter, Visual Studio has many templates that we can use with .NET and other technologies, depending on the type of project you are working on.

To explore the templates in Visual Studio 2022, just open Visual Studio and select the **Create a new project** option:

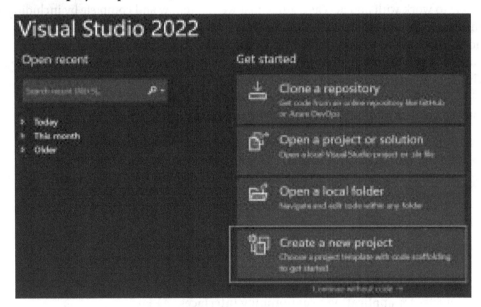

Figure 4.1 – The Visual Studio Create a new project option

After selecting this option, we will see a list of templates to choose from for creating our new project, as shown in *Figure 4.2*. Similarly, on the left side, we will be able to see the templates we have recently selected, with the option to pin or unpin each template, so that we can quickly select the templates that we use the most:

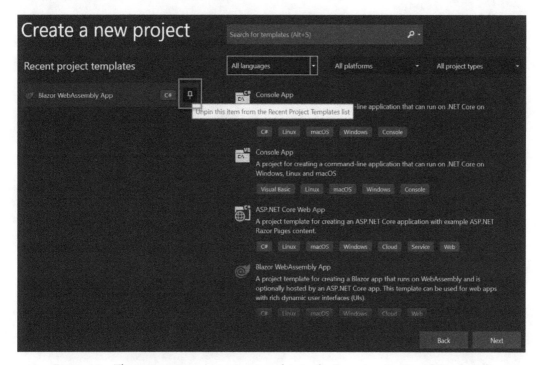

Figure 4.2 – The option to pin or unpin a template in the Recent project templates section

In *Figure 4.2*, we also see a search bar at the top for filtering the different templates. If you know the name or the technology that you will use, you can type the term related to it and start searching for the template. The search engine will show you all the templates that include the term you have entered.

Below the search bar, we can also see three drop-down controls, which also serve as filters to search for templates related to the programming language, platforms, or types of projects we want to create. Among the most important options for each of these filters, we can find the following:

- **All languages**:

  - **C#**
  - **JavaScript**
  - **Python**
  - **TypeScript**
  - **Visual Basic**

- **All platforms**:

  - **Android**
  - **Azure**
  - **iOS**
  - **Linux**
  - **Windows**

- **All project types**:

  - **Cloud**
  - **Desktop**
  - **Games**
  - **Machine Learning**
  - **Mobile**
  - **Test**
  - **Web**

In addition, it is possible to combine the filters with the search bar. For example, you can type .NET Core and select the **C#** option in the language drop-down menu to get all the projects related to .NET Core and C#, as shown here:

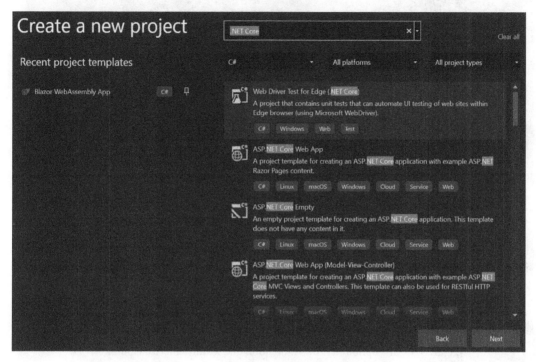

Figure 4.3 – The .NET Core templates for C#

In *Figure 4.3*, we can see many kinds of projects related to .NET Core that Visual Studio has found, according to the filters and workloads selected during installation.

At the end of the results, you will have an option to install other templates if you cannot find the option that you are looking for:

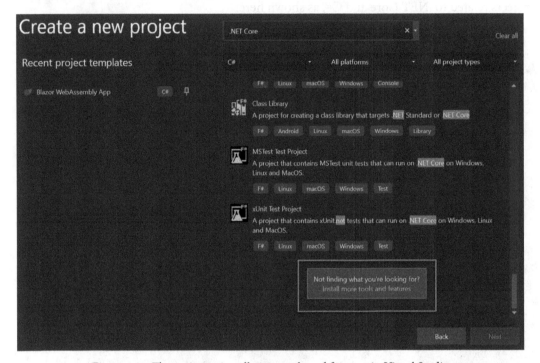

Figure 4.4 – The option to install more tools and features in Visual Studio

It is important to note that each template has a description including the details related to the template, so we can easily identify whether the template includes the structure and schema that we need.

Lastly, when the description of the template contains the **Empty** keyword, it means that the template doesn't contain elements or modules by default and includes either demos or examples, such as a guide, for the developer. This kind of template only includes the project and the base components to compile and run.

Now that we know how to use the template selection window, let's look at the most important templates for web development.

Templates for .NET Core

Let's start this section knowing the difference between .NET Core and ASP.NET Core:

- .NET Core (now called .NET) is a cross-platform, free, and open source framework used to create web, desktop, mobile, and other kinds of applications, using a standard library and C#, Visual Basic, or F# as a programming language.

- ASP.NET Core, on the other hand, is a multi-platform web technology for creating modern applications using .NET. This is why, although the .NET Core framework has been replaced with .NET 6, you will still find terms related to .NET Core as a part of the templates.

If we want to create a new clean project of the ASP.NET Core type, we can search and select the **ASP.NET Core Empty** template. This template will create a simple API project that, when executed, will display the **Hello World** message. To configure the new project, we must enter a project name, such as `AspCoreEmpty`.

Also, we must indicate the location to save the project. You can choose to leave the default path, or select a custom path, such as `C:\demos\AspCoreEmpty`.

Finally, set the name of the solution. By default, it will be the same as the project, but you can change it to another name. In *Figure 4.5*, we have kept the same name:

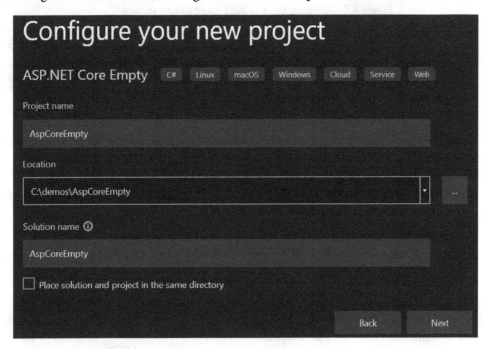

Figure 4.5 – Configuring a new ASP.NET Core Empty project

After clicking on the **Next** button, you will be shown a new window, as follows:

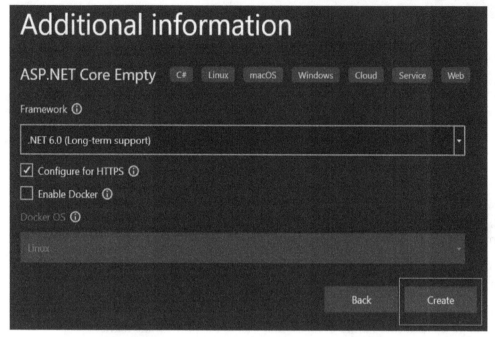

Figure 4.6 – Creating the ASP.NET Core Empty project

As shown in *Figure 4.6*, you can configure the following options:

- **Framework**: This dropdown will show you the set of frameworks available to work with the selected technology. Although it is possible to select older frameworks, it is always advisable to create new projects with the latest version of the framework available.

- **Configure for HTTPS**: This checkbox allows you to configure the project to use a self-signed SSL certificate. If you are working on such a project for the first time, you will be asked to trust the certificate when you run the application so that everything works correctly. Although it is possible to work with the HTTP protocol, it is always recommended to use HTTPS in real life.

- **Enable Docker**: This option allows you to enable Docker support in your project. This means that a Docker file will be generated, which you can then publish – for example, to Microsoft Azure.

In our example (*Figure 4.6*), we select **.NET 6** as a target framework. The **Configure for HTTPS** option is marked by default, and it is optional for this demo. Finally, you can click on **Create** to complete the wizard and create the project.

After creating the project, we can analyze the structure of the created project, which is simple, as shown in *Figure 4.7*:

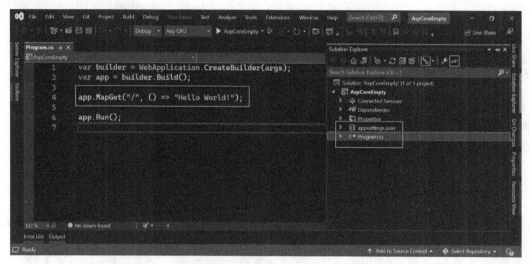

Figure 4.7 – The ASP.NET Core Empty project in Visual Studio

In the panel called **Solution Explorer**, we can see all the files related to the template created for the project.

There are only two files related to the structure of the project, which are the following:

- `appsettings.json`: This is a JSON file that contains all the settings by default for an ASP.NET Core project.

- `Program.cs`: This is the entry point of an application that contains a sequence of commands to configure and start the application.

In this template we have selected, a new concept introduced in .NET 6 is used, called Minimal APIs, which is a way to have the minimum code necessary to create a simple endpoint, using helper functions inside `Program.cs`.

By default, when the application is executed, it will show a **Hello world** message, returned in a GET method and mapped into the base URL – for example, `http://myapiurl/`.

> **Important Note**
> Minimal API is a new template included in .NET 6 that is very useful for creating simple APIs, such as demos, small services, serverless functions, and microservices.

Now, let's create a new project with the **ASP.NET Core Web App** template to analyze its structure. Open Visual Studio 2022 as we did in *Figure 4.4*, and select **Create a new project**. Then, search for the term ASPNET and select the **ASP.NET Core Web App** template:

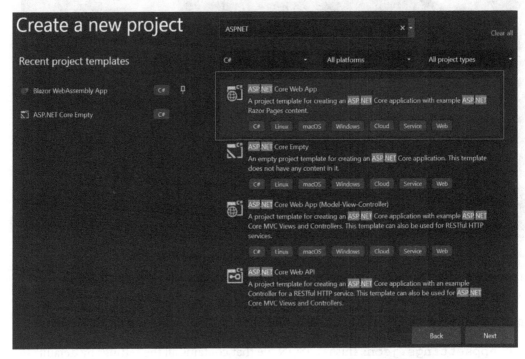

Figure 4.8 – Searching by ASPNET and selecting the ASP.NET Core Web App template

Then, click on **Next** and fill out the information required for the new project, as shown in *Figure 4.9*:

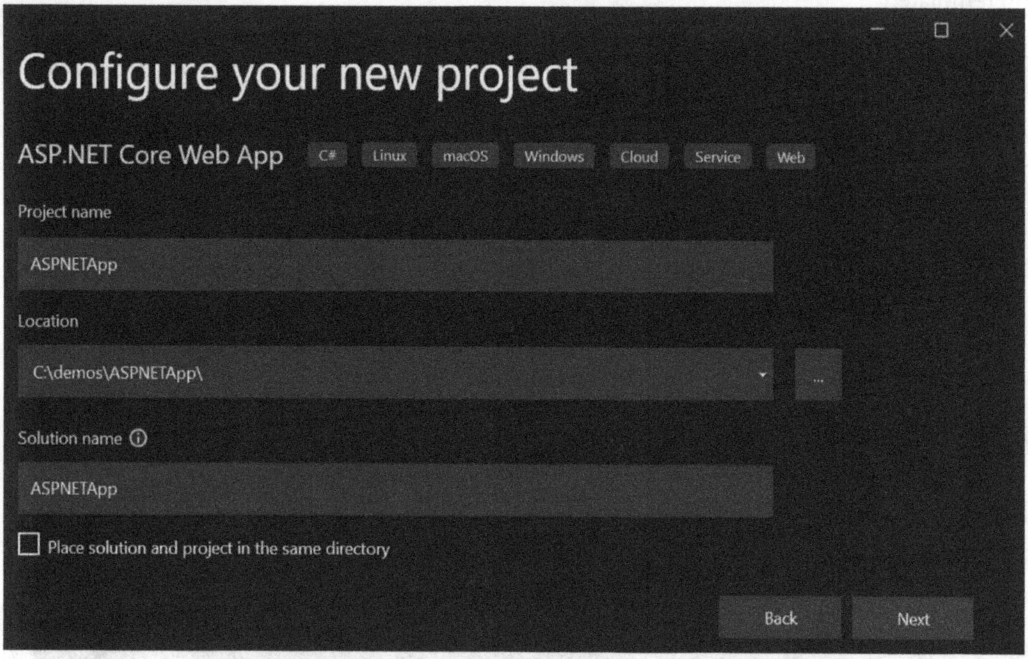

Figure 4.9 – Configuring the new ASP.NET Core Web App project

After filling out the additional information for the project, as seen in *Figure 4.10,* you can click on the **Create** button to generate the project with the ASP.NET Core Web App template:

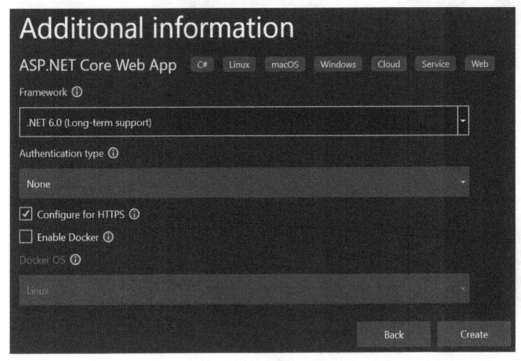

Figure 4.10 – Additional options for the new ASP.NET Core Web App template

This template is perfect if you want to create a web application using .NET and C# running on the server. Also, it uses razor pages (syntaxes that combine C# with HTML in cshtml extension files) to build a web application into small and reusable pieces.

Modern applications normally run in the browser because they run faster and have a better **search engine optimization** (**SEO**), which improves the traffic to your website. However, server-side applications are still very useful for dashboards, internal projects, administration panels, and many other types of web applications.

An ASP.NET Core Web project contains a wwwroot folder, which you can see in *Figure 4.11*:

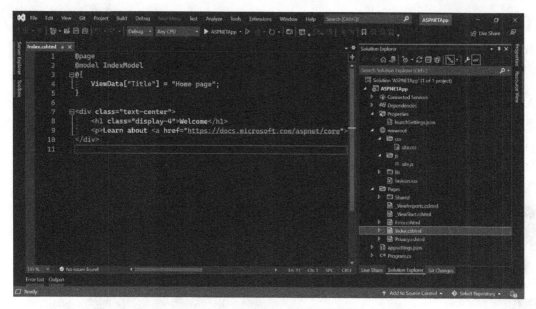

Figure 4.11 – The structure of an ASP.NET Web project

The `wwwroot` folder is associated with static files such as CSS, images, and JavaScript files. There is also a folder called `Pages` that contains all the UI pieces.

`Index.cshtml` is an example where you can see C# code mixed with HTML code. The @ character allows you to use the C# code in the file. `@model`, for example, sets the model type to map the values on the page.

Other kinds of .NET Core project templates include the following:

- **Console App:** This template creates a simple console application. Normally, we use this kind of application when we are starting out and learning.

- **Class Library:** This allows you to create components and classes to share across different projects.

- **Templates for APIs**

- **Templates for SPAs**

These are the recommended templates, as they are the most updated and in line with the latest versions of .NET.

Templates for APIs

Today, it is almost a rule that applications use API-based endpoints. This is because they provide a high level of security and interoperability by not depending on a particular technology or operating system, and they can be scaled according to the existing demand. Also, we can implement advanced architecture, such as microservices using APIs, where all our business logic is distributed in small isolated and standalone services.

Due to the great importance of being able to develop solutions based on APIs, Visual Studio 2022 incorporates a special template so that we can create APIs based on .NET 6, which we can search by filtering the `api` term, as can be seen in *Figure 4.12*:

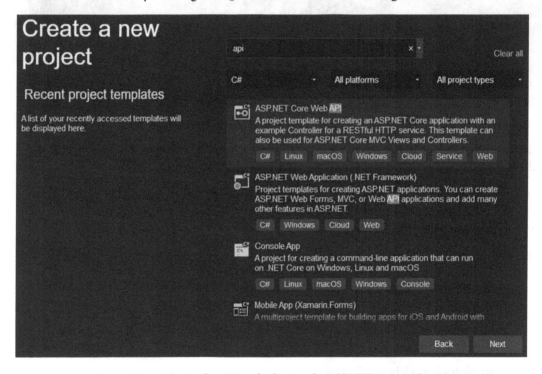

Figure 4.12 – Filtering by API and selecting the ASP.NET Core Web API

Once we have selected the template, we will have to fill in the complementary information, according to our needs, as seen in *Figure 4.13*:

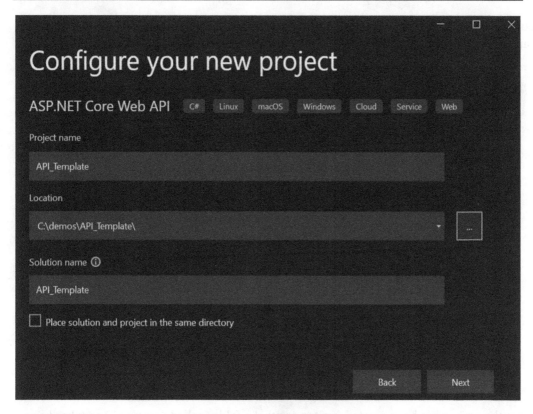

Figure 4.13 – Configuring the ASP.NET Core Web API project

In the next window, we will be asked to select the version of the framework and whether we require some type of authentication, among other data that we have already seen in the *Templates for .NET Core* section. However, we have a couple of additional options that we do not see in a normal ASP.NET Core project, which are as follows:

- **User controllers**: If this option is selected, the use of controller files will be enabled from the configuration. If it is deselected, a feature called minimal APIs will be used, which will create the minimum code necessary to have a functional API.

- **Enable OpenAPI support**: Swagger is a set of open source tools based on the OpenAPI specification that will allow us to describe the APIs we create in a simple, easy, and well-structured way, providing API users with good documentation.

For our example, we will leave both options selected, as seen in *Figure 4.14*, so that you can see the resulting structure:

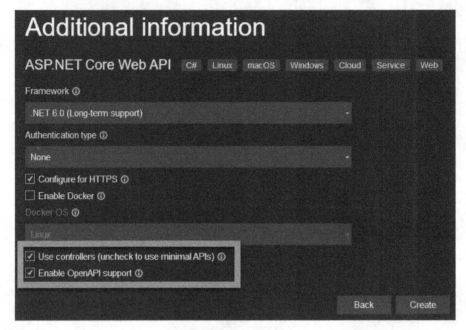

Figure 4.14 – Selecting options in Additional information for the ASP.NET Core Web API project

Once the project has been created, we can see that a folder called `Controllers` has been created as part of the project:

Figure 4.15 – The structure of an ASP.NET Core API project

In the `Controllers` folder, we will find the controllers that will be created as part of the API. Remember that we can see the controllers in this folder because of the **Use controllers** option we selected in *Figure 4.14*.

To run or execute the application, we need to use the green arrow or play icon in the standard toolbar:

Figure 4.16 – The play button to start the project in Visual Studio

We can start the project with debugging on or off. In this case, either option is fine. We just want to see how the project looks.

> **Important Note**
>
> There are some useful shortcuts to start a project. You can use *F5* on your keyboard to run the project in debugging mode, and *Ctrl + F5* to start the project without debugging.

If we proceed to run the application, it will immediately take us to a window with the URL ending at `/swagger/index.html`:

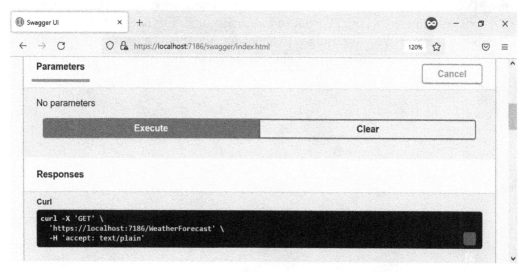

Figure 4.17 – The Swagger page for the ASP.NET Web API project

Here, we will be able to find all the functionality available in our API documented through Swagger – this means all the endpoints that we create from the controllers in our project. Swagger is a library that uses the OpenAPI standard to show all the endpoints, including the parameters required and the type of values returned.

From this same page, we will be able to test each one of the endpoints, for the purpose of validating them, and carry out necessary debugging when some endpoint does not work as expected. As you can appreciate, this tool is of great help, both for development purposes and for providing users with documentation of the created API.

This template gives us everything we need to create our own modern APIs from scratch. Let's now take a look at the templates available for working with .NET Framework.

Templates for .NET Framework

.NET Framework is the original implementation of .NET, released by Microsoft in 2002. The long-term idea was to make this framework a multi-platform framework. Unfortunately, it always worked officially on Windows devices, due to API restrictions. To solve this problem, specific versions of the framework were created to support different platforms, which caused a fragmentation of the platform. Through the experience gathered from these projects launched over time, the Microsoft team has finally succeeded in unifying the .NET platform.

As, for many years, the .NET Framework platform was the primary framework for developing .NET applications, many companies around the world use ASP.NET Web Forms, MVC, or Web API applications, based on .NET Framework. This is the reason why .NET Framework templates are still included in Visual Studio. This does not mean that it is advisable to create applications using such templates. You should always opt for the use of modern technologies that have current and constant support.

Although there are several templates with .NET Framework, there is one that really interests us for web development. This is called **ASP.NET Web Application (.NET Framework)**, which you can search for with the net framework term, as seen in *Figure 4.18*:

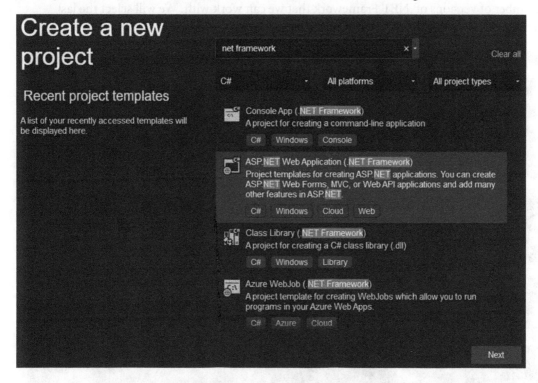

Figure 4.18 – Searching with net framework

Once you select the template, in the next window, you will see numerous fields to fill in, which will be familiar to you, such as the project name and location, among other data. The important point is to note that we have a dropdown that shows us a significant number of versions of .NET Framework that we can work with. We will select the last version in this example, as you can see in *Figure 4.19*:

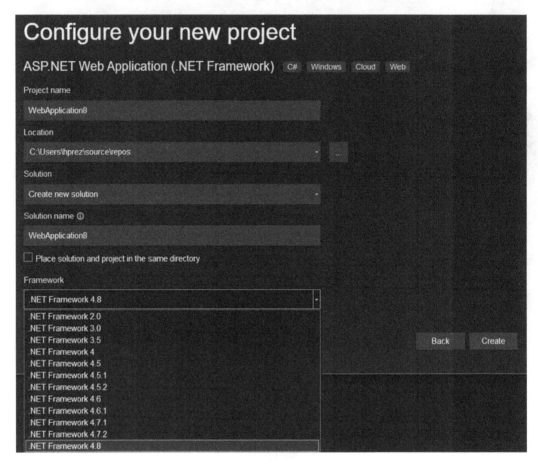

Figure 4.19 – Configuring ASP.NET Web Application (.NET Framework)

Once we have selected the framework, we can choose what type of application we want to create – the options include a clean project, templates for web forms, MVC-based projects, web APIs, and single-page applications, as seen in *Figure 4.20*:

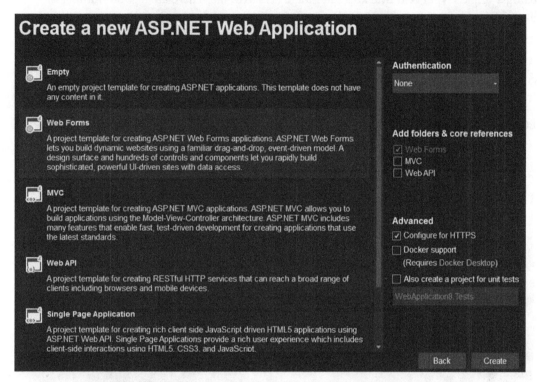

Figure 4.20 – The type of web applications for .NET Framework in Visual Studio

The selection of these templates will depend on the needs of the application, although, as mentioned at the beginning of this section, it is recommended to opt for templates that are oriented toward .NET.

Let's move on to review the category of SPA's templates, which is oriented toward the creation of applications with which users will interact.

Templates for SPAs

SPAs is an amazing architecture for web projects, where all the elements are rendered using a single **HyperText Markup Language** (**HTML**) file. There are a good number of libraries and frameworks that use this concept – for example, Angular, React.js, and Blazor WebAssembly.

Any project created with these templates will contain all the required components to create a monolithic application, using ASP.NET on the backend and a SPA library or framework on the frontend side.

Let's create our first SPA project with Visual Studio. You can search for `ASP.NET Core with` to find the templates for SPAs:

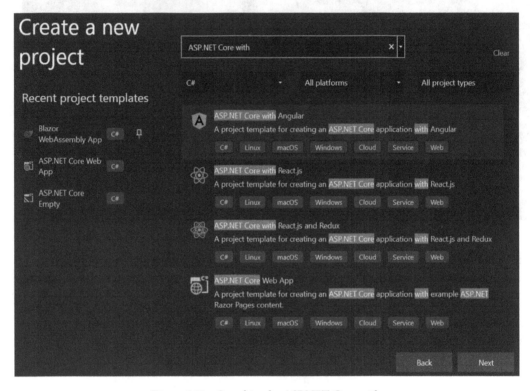

Figure 4.21 – Searching by ASP.NET Core with

In *Figure 4.21*, you can see the result of searching by ASP.NET Core with. There are three templates that we can select to work with SPAs:

- **ASP.NET Core with Angular**
- **ASP.NET Core with React.js**
- **ASP.NET Core with React.js and Redux**

We will create a project with React.js, so let's select **ASP.NET Core with React.js** and then click on **Next** to continue:

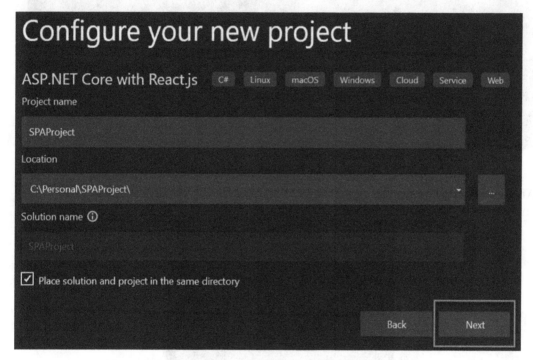

Figure 4.22 – Configuring ASP.NET Core with the React.js project

---

Important Note

This project that you will create is going to be used for the next exercises in this book, so it's important to save this code in a permanent folder that you prefer.

Then, you should select **.NET 6 (Long-term support)** as a **Framework** target and **None** in **Authentication type**, and finally, click on **Create**, as shown in *Figure 4.23*:

Figure 4.23 – Filling out the Additional information section

After clicking on **Create**, you can see the project created with the SPA template and analyze the architecture:

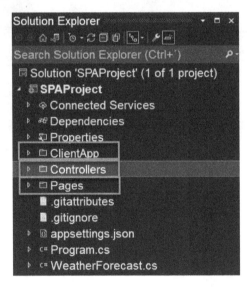

Figure 4.24 – The ASP.NET Core project with the React.js project created in Visual Studio

There are three important folders to highlight in this template:

- **ClientApp**: Contains the client application – in this case, a React.js app

- **Controllers**: Contains all the controllers related to the business logic on a server

- **Pages**: Contains razor pages, which means UI components rendered on a server

The template has a demo with the `WeatherForecastController.cs` file. This is a simple demo that returns some random data.

Let's run the project by clicking the play button, as shown in *Figure 4.25*, to see how the application looks:

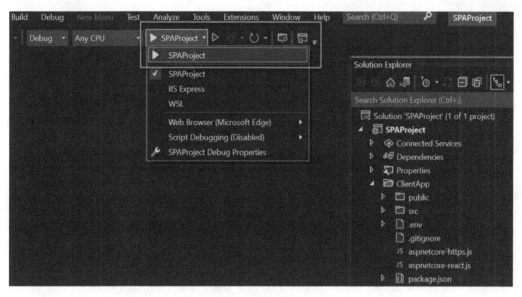

Figure 4.25 – The option to run the project in Visual Studio (the green arrow or play icon)

When the project is running, you can see a simple HTML page explaining how the template works, which includes two demos, **Counter** and **Fetch data**:

SPAProject   Home Counter Fetch data

Hello, world!

Welcome to your new single-page application, built with:

- ASP.NET Core and C# for cross-platform server-side code
- React for client-side code
- Bootstrap for layout and styling

To help you get started, we have also set up:

- **Client-side navigation**. For example, click *Counter* then *Back* to return here.
- **Development server integration**. In development mode, the development server from `create-react-app` runs in the background automatically, so your client-side resources are dynamically built on demand and the page refreshes when you modify any file.
- **Efficient production builds**. In production mode, development-time features are disabled, and your `dotnet publish` configuration produces minified, efficiently bundled JavaScript files.

The `ClientApp` subdirectory is a standard React application based on the `create-react-app` template. If you open a command prompt in that directory, you can run `npm` commands such as `npm test` or `npm install`.

Figure 4.26 – ASP.NET Core with the React.js project running

The template is simple but includes everything that we need for creating a new web application using a monolithic architecture and best practices on the backend and frontend sides. If you require a web application with high-performance security, this template is a good option. React.js uses JavaScript in the syntaxes, so for this template, you need more knowledge of this language and C# for the backend. In the template that we analyzed in the *Templates for .NET Core* section, C# was more important than JavaScript for writing our UI and business logic.

We will perform some modifications in this project, and we will use the files and demos by default to analyze the tools in Visual Studio and learn how to take advantage of these amazing functionalities.

Summary

Visual Studio has different options for creating projects, using templates depending on the workload added during the installation process. We have four filters to search quickly in the templates. Though we can filter by language, platform, and project type, we also have the possibility to use the search bar to find our template, using a term or a specific word.

After selecting the template to use, we must always type a project name and the folder where our project is going to be created. Then, we must select the target framework and some optional information. Visual Studio is going to open the project after selecting **Create** on the screen, so we have the possibility to start working on our project right away.

Using **Solution Explorer**, we can see the structure of the project, including the folders used in the application. Normally, a project created with a template contains a demo that helps us to verify whether an application is running fine.

In *Chapter 5, Debugging and Compiling Your Projects*, we will debug our created SPA project. We will understand why this tool is important, and we will check the options to compile our project before executing it.

5
Debugging and Compiling Your Projects

As software developers, one skill that should be learned as early as possible is program debugging. This applies to .NET projects but also any other software development technology.

If you want to get the most out of Visual Studio 2022, you must be familiar with its different windows that can help you observe information to fix bugs and know how to use as many of the debugging tools it offers, including **breakpoints**. A breakpoint offers the functionality to stop the execution of an application, allowing you to see the state of each of the objects and corroborate its behavior.

That is why, in this chapter, we will talk about debugging in Visual Studio 2022, the different breakpoints available, the state inspection tools, and which are the best scenarios for each one of them.

The topics we will discuss in this chapter are as follows:

- Debugging projects in Visual Studio
- Exploring breakpoints in Visual Studio
- Inspection tools for debugging

Let's learn about debugging in Visual Studio 2022 to detect possible errors in our programs.

Technical requirements

To follow the examples shown in this chapter, Visual Studio 2022 must be installed with the web development workload, as shown in *Chapter 1, Getting Started with Visual Studio 2022*.

Likewise, the code implementation will be performed on the project created in *Chapter 4, Creating Projects and Templates*, specifically in the `Program.cs` file, for easier testing.

You can find the `Program.cs` file with the changes made throughout this chapter at the following link: `https://github.com/PacktPublishing/Hands-On-Visual-Studio-2022/blob/main/Chapter05/Program.cs`.

Debugging projects in Visual Studio

Before discussing the breakpoint topic in depth in the *Exploring breakpoints in Visual Studio 2022* section, it is important that you know some technical aspects used in the debugging world, as well as in Visual Studio.

Understanding the technical aspects of debugging

It is important that you know the difference between the terms debugger and debugging so that you know what I mean when I mention any of these terms during the chapter and the book.

First, the term debugging refers to the action of looking for errors in the code. This does not necessarily include the use of a tool such as an IDE. You could, for example, search for errors in code written on a piece of paper, and you would still be debugging.

This is usually not feasible, and a tool called a debugger is often used. This tool is attached to the application process you are going to run, allowing you to analyze your code while the application is running.

Differentiating between debug mode and run mode

It is essential to differentiate between debug mode and run mode in Visual Studio, as they can be confusing to those who touch the IDE for the first time.

Let's first analyze debug mode. This option is activated by selecting the **Debug** configuration (which is preselected by default) and clicking on the green button located in the same space as the project name, as shown in *Figure 5.1*:

Figure 5.1 – Visual Studio's debug mode option

When this option is pressed, the debugger will be attached to the execution of the application, which will allow us to use functions, such as stopping at a certain breakpoint in our application.

On the other hand, we can also choose a second configuration in the drop-down list, as shown in *Figure 5.2*, called **Release**:

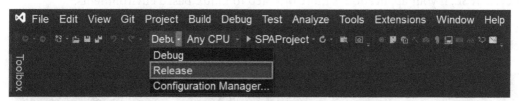

Figure 5.2 – Visual Studio's release mode option

When this option is selected and we proceed to start the execution of the application, the debugger will not be attached, which will give you a better idea of how your application will behave toward the end user. This implies that you will not be able to perform code debugging or see where exceptions have occurred, but you will gain a performance improvement.

Project debugging initialization options

As part of the debugging and execution options of an application, we must know that we have a set of options available to perform our tests.

If you drop down the options next to the green button with the name of your project, as shown in *Figure 5.3*, you will be able to see a set of configurations for the deployment of your application:

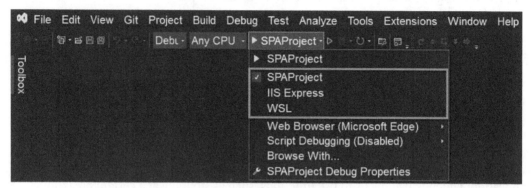

Figure 5.3 – The server configuration options for debugging

For example, the first three options are used to choose which server you want to use for testing. By default, a server called **Kestrel** is used. But it is also possible to use **IIS Express**, or even **WSL**, if your application is more oriented to Linux-based environments.

In *Figure 5.4*, we can observe the following options to change the browser, enable script debugging, and view the debugging properties of the project:

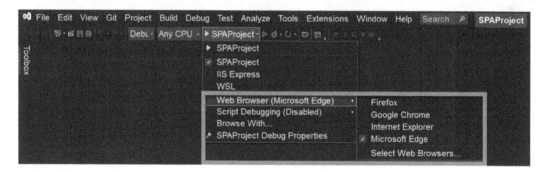

Figure 5.4 – Additional options for debugging

These options are useful if, for example, you want to use a particular browser to use some of its own tools.

If the application has been configured in debug mode, you will see the buttons that allow you to control the execution of the application in the upper part of the IDE, as shown in *Figure 5.5*. These buttons, from left to right, are used to do the following:

- Pause the application

- Stop the application
- Restart the application

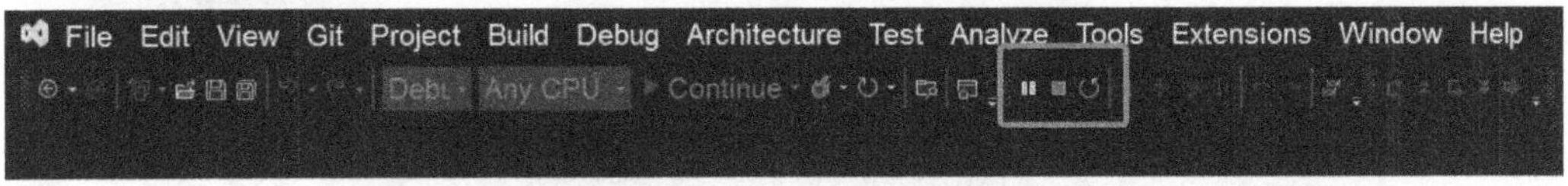

Figure 5.5 – Buttons to control the flow of the application

Now that we know the existing debugging concepts in Visual Studio, let's analyze breakpoints.

Exploring breakpoints in Visual Studio

Breakpoints are a fundamental part of software development. They allow you to stop the flow of your application at any point you want to inspect the state of your objects.

To place a breakpoint in Visual Studio, it is enough that we position ourselves right next to the numbering of the lines. This will start showing a gray circle that will appear and disappear as we move the cursor over the line numbers, as shown here:

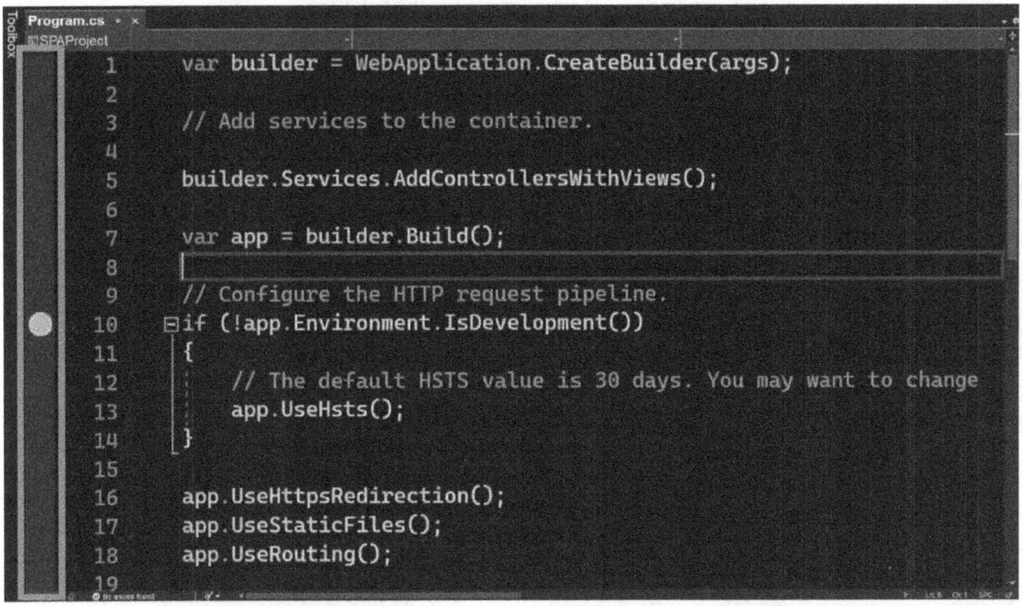

```csharp
1   var builder = WebApplication.CreateBuilder(args);
2
3   // Add services to the container.
4
5   builder.Services.AddControllersWithViews();
6
7   var app = builder.Build();
8
9   // Configure the HTTP request pipeline.
10  if (!app.Environment.IsDevelopment())
11  {
12      // The default HSTS value is 30 days. You may want to change
13      app.UseHsts();
14  }
15
16  app.UseHttpsRedirection();
17  app.UseStaticFiles();
18  app.UseRouting();
19
```

Figure 5.6 – The location for placing breakpoints in Visual Studio

Once we find the line we want to debug, we just need to left-click once, which will cause the circle to turn red, as shown in *Figure 5.7*. Once you have done this, if you move the cursor away from the circle, you will be able to see how it remains as it is, as shown here:

```
 5        builder.Services.AddControllersWithViews();
 6
 7 ⚲    var app = builder.Build();
 8
 9        // Configure the HTTP request pipeline.
10      ⊟if (!app.Environment.IsDevelopment())
```

Figure 5.7 – Placing a breakpoint

If we proceed to execute the application with the breakpoint set, we will see how the application flow stops immediately after starting the application, as shown in *Figure 5.8*:

```
 5        builder.Services.AddControllersWithViews();
 6
 7 ⚲    var app = builder.Build();
 8
 9        // Configure the HTTP request pipeline.
10      ⊟if (!app.Environment.IsDevelopment())
```

Figure 5.8 – Debugging a breakpoint

Once the application has stopped at the breakpoint, we have different tools available that we can use to view the status of the application – for example, if we position over a variable that is before the debug line, we will be able to see its current status. If it is primitive data, you will see its value immediately, while if it is an object, you will be able to go inside its properties to examine each one of them, as shown in *Figure 5.9*:

Figure 5.9 – Examining the properties of an object

There are more options to place breakpoints – for example, if you place yourself on one of the gray circles and click on it with the right button, a series of breakpoints will appear, which we will analyze in the *Conditional breakpoints*, *Temporary breakpoints*, and *Dependent breakpoints* sections so that you can insert them:

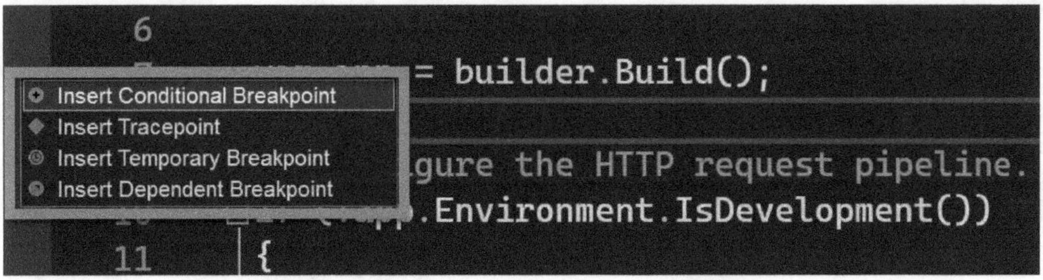

Figure 5.10 – The options for inserting a breakpoint

Likewise, if you right-click on any breakpoint already placed, you will see options to add functionalities to them:

Figure 5.11 – The menu for modifying breakpoints

This is the easiest way to add breakpoints to your project. However, you will often need special breakpoints that are activated under certain circumstances, so we will proceed to review them in the next sections.

> **Important Note**
>
> Visual Studio has keyboard shortcuts created for almost all the operations we are going to be performing. So, I will mention them for you as *notes*, as we move forward throughout this chapter.
>
> It is possible to enter a breakpoint quickly by placing yourself on the line you want to debug and pressing the *F9* key.

Navigating between breakpoints

Once we know how to place breakpoints in the source code, we can continue the execution of the application in different ways through the buttons located in the upper part of the menu:

Figure 5.12 – The debugging options for application execution

Each of them will execute the code as follows:

- **Step into**: This allows us to execute all the source code line by line. If, as part of the flow, we invoke methods to which we have access through the source code, we will navigate to it automatically following the line-by-line debugging.

- **Step over**: This allows us to only walk through lines of code in the current method and not step into any methods invoked by the current method.

- **Step out**: This button is used in case we are inside a method. It will allow us to step out of the execution of the method to return just to the line after the invocation of the method.

> **Important Note**
> Each of these options has a shortcut assigned to it:
> Step into: the *F11* key
> Step over: the *F10* key
> Step out: *Shift + F11* keys

If we wish to examine all the breakpoints we have in our project, we can do it by activating the breakpoints window. This is done from the **Debug | Windows | Breakpoints** menu, as seen in *Figure 5.13*:

Figure 5.13 – The option to display the breakpoint window

This will display a new window that shows a list of all breakpoints placed in our project.

Conditional breakpoints

There are times when you will need your breakpoint to stop when certain conditions are met. In this case, using conditional breakpoints is the best option. To insert a breakpoint of this type, just right-click on the sidebar, which will show you the different types of breakpoints available, as shown in the *Navigating between breakpoints* section. Select the **Conditional Breakpoint** type, which will open a window with preselected options, as shown in *Figure 5.14*:

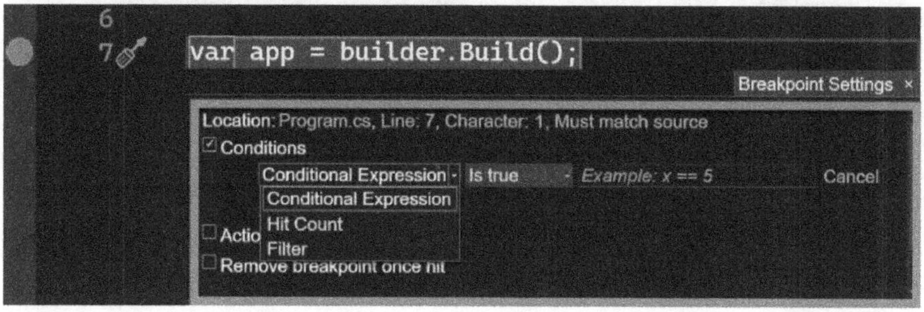

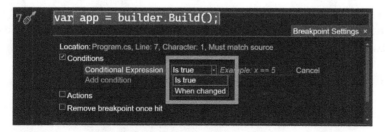

Figure 5.14 – The options for conditional breakpoints

Within a conditional breakpoint, we can configure either a **Conditional Expression** condition, a **Hit Count** condition, or a **Filter** condition, which we will discuss in the next sections.

> **Important Note**
>
> To insert a conditional expression, you can use the *Alt + F9* keys, followed by the *C* key.

Conditional Expression

The **Conditional Expression** option, as shown in *Figure 5.15*, will allow us to stop the application when a true condition that we have previously specified is fulfilled, or when the value of an object changes:

Figure 5.15 – The types of expression that can be evaluated with conditional expressions

In the first case, we can specify a Boolean expression, ranging from the comparison of a pair of values to the comparison of complex objects.

In *Figure 5.17*, a `for` loop has been created that prints the numbers from 1 to 10 to have a better appreciation of the example. Also, I have placed a pin on the i variable, which you can place by simply hovering over any variable while the application is running and clicking on the pin symbol, as shown here:

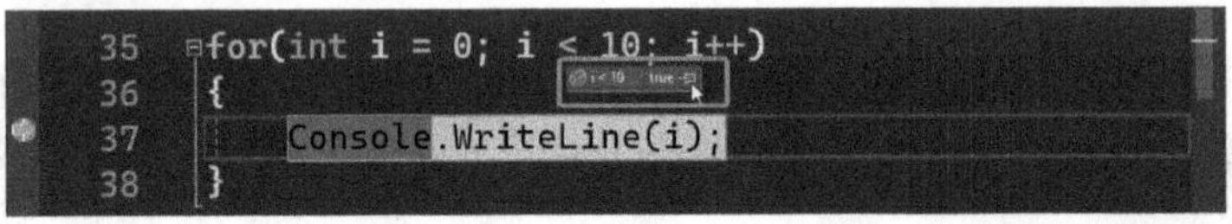

```
35    for(int i = 0; i < 10; i++)
36    {
37        Console.WriteLine(i);
38    }
```

Figure 5.16 – The types of expressions that can be evaluated with conditional expressions

As a condition, we have indicated that we only want to stop the application when the value of i is greater than the number 5. After starting the application, the breakpoint has stopped, just when i has a value of 6:

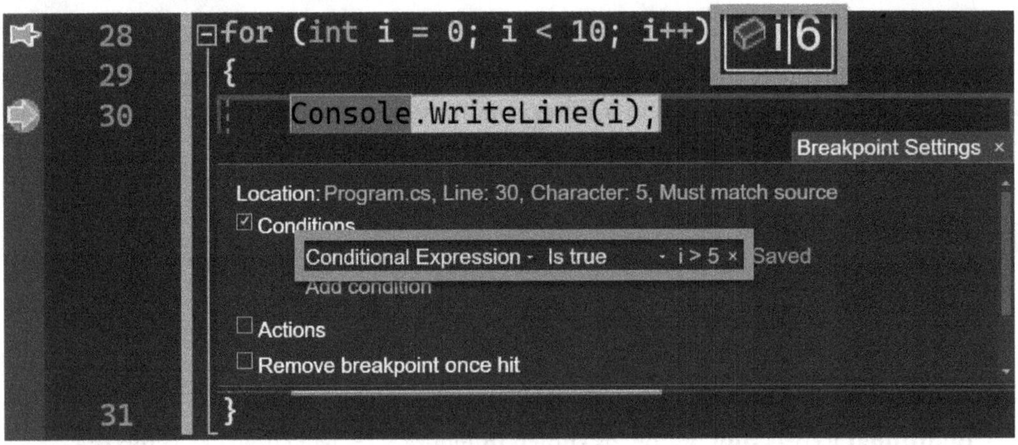

```
28    for (int i = 0; i < 10; i++)      i 6
29    {
30        Console.WriteLine(i);
                                         Breakpoint Settings ×

      Location: Program.cs, Line: 30, Character: 5, Must match source
      ☑ Conditions
          Conditional Expression - Is true        · i > 5 × Saved
          Add condition

      ☐ Actions
      ☐ Remove breakpoint once hit

31    }
```

Figure 5.17 – Debugging a conditional expression

In the second case, we can monitor an object, property, or field so that the breakpoint stops only when it changes. This can be seen in a practical way in *Figure 5.18*, in which the **Conditional Expression** type has been changed to **When changed**, and it has been indicated that we want to monitor the i variable:

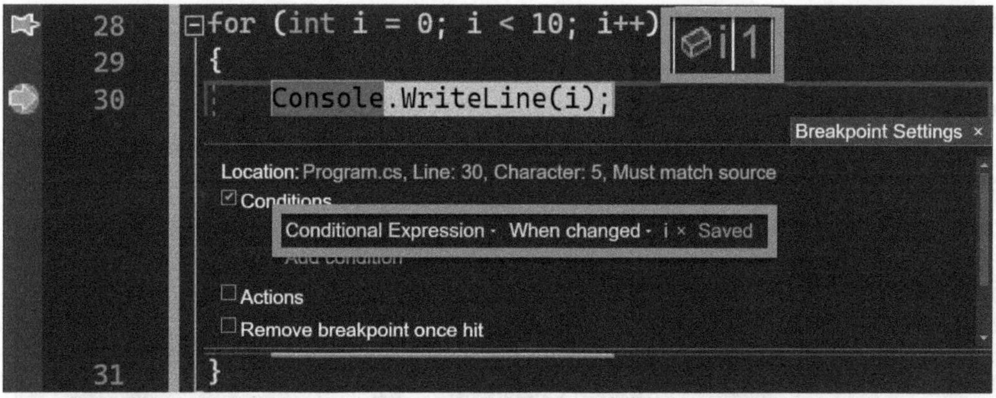

Figure 5.18 – Evaluating a conditional expression with the When changed parameter

As a result, when the breakpoint is executed for the first time, the value of i is equal to 1.

Hit Count

The **Hit Count** condition type is intended to solve problems in loops – for example, suppose you must print a certain number of reports according to a set of conditions but suddenly realize that after report 25, the calculations start to go wrong. The first thing you might think of doing is to set a breakpoint, start the application, run the code 25 times to get to iteration 25, and evaluate it. Although this is the correct way to do it, a more optimized way is to set the **Hit Count** option and specify the number of iterations you want to skip – in this example, 25, as we can see in *Figure 5.19*:

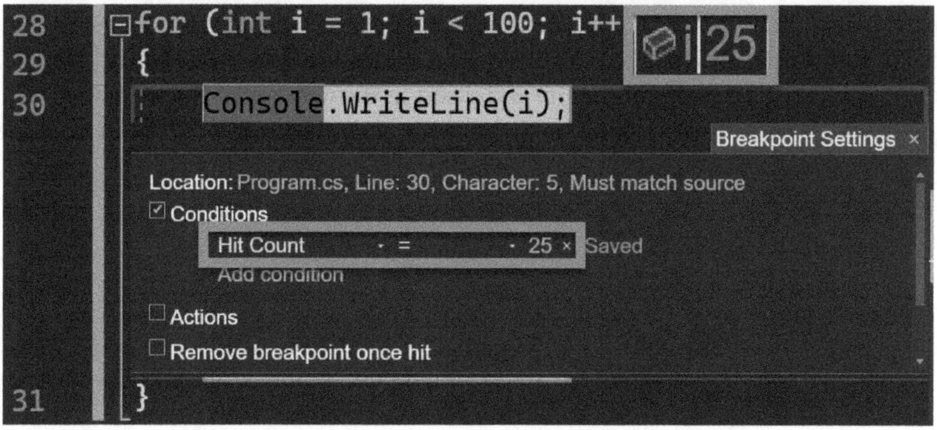

Figure 5.19 – Evaluating a Hit Count expression

This will cause the application to stop when i has a value of exactly 25.

Filter

The last type of conditional breakpoint is the **Filter** condition. This will allow us to trigger a breakpoint according to a series of predefined expressions. These expressions can be found in the **Filter** box, as shown in *Figure 5.20*, and range from the machine name, through processes, to thread properties:

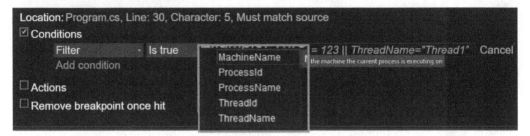

Figure 5.20 – The options for a Filter condition

These are the types of conditional breakpoints, which are very useful for performing value comparisons. Now, let's look at function breakpoints.

Function breakpoints

The **Function Breakpoint** type, as its name indicates, will allow us to debug a method when it is executed, even if we have not set a breakpoint as we did in the *Conditional breakpoints* section. This is very useful if you have hundreds of lines of code and know the name of the function that you want to debug.

Unlike the breakpoints we saw in the *Conditional breakpoints* section, this breakpoint is placed in a different way. First, as seen in *Figure 5.21*, you can go to the **Debug | New Breakpoint | Function Breakpoint** menu and insert in the window the name of the function in which we want to set the breakpoint, instead of placing the red dot in the code:

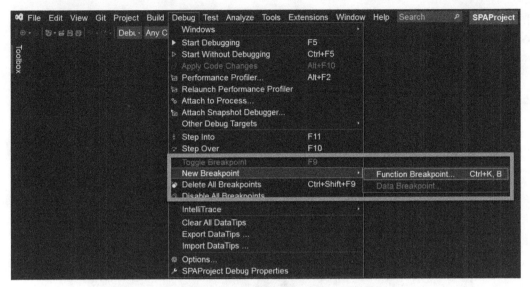

Figure 5.21 – Adding a function breakpoint from the Debug menu

The next way is to add a breakpoint function from the **Breakpoints** window, as shown in *Figure 5.22*:

Figure 5.22 – Adding a function breakpoint from the Debug menu

Once we click on the **Function Breakpoint** button, a new window will pop up, asking for the name of the method we want to monitor. For the purposes of this demonstration, we have introduced a very simple function to the end of the `Program.cs` file, which has the following format:

```
void StopHere()
{
    Console.WriteLine("Hi!");
}
```

If we know the name of the function to be evaluated, we can enter it in the **Function Name** box. We can do this in several ways:

- Typing the name of the function
- Specifying the function name with a specific overload
- Specifying the `dll` name if we have the source code for it

In our example, we will place only the name of the `StopHere` function:

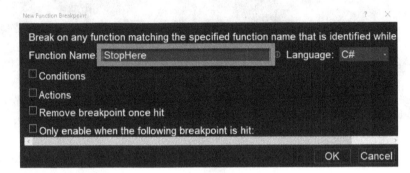

Figure 5.23 – Filling in the name of the function we are interested in debugging

With the data established, we will proceed to execute the application, having previously called this new method. This will cause the application to stop at the start of the method (which we specified in *Figure 5.23*), as seen in *Figure 5.24*:

```
26
27    StopHere();
28
29    app.Run();
30
31    void StopHere()
32    {
33        Console.WriteLine("Hi!");
34    }
```

Figure 5.24 – A demonstration of the breakpoint being executed through a function breakpoint

---

Important Note

It is possible to add a function breakpoint by pressing the *Ctrl* + *K* keys, followed by the *B* key.

Undoubtedly, this type of breakpoint will make our life easier when we want to debug methods quickly. In case you want to trigger breakpoints based on the data of an object, you can use the data breakpoints, which we will see next.

Data breakpoints

If you want to be able to place breakpoints when the properties of an object change, then data breakpoints are your best option. If you try to add one of these breakpoints from the **Breakpoints** window, you will see that the option is disabled.

This is because we first need to place a breakpoint at a point before we want to start monitoring the property. Once this is done, start the application until the breakpoint is activated, and in one of the windows called **Autos**, **Watch**, or **Locals**, where the instance that interests you appears, you must right-click to see the option called **Break When Value Changes**, as shown in *Figure 5.25*, in which we want to monitor the EnvironmentName property of the app object:

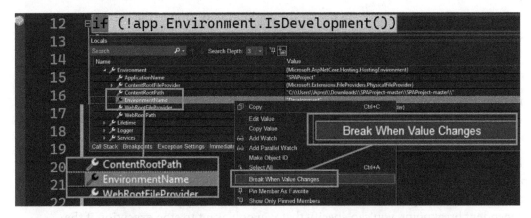

Figure 5.25 – Adding a data breakpoint to break when the EnvironmentName property changes

Once we select the option, we will see how a new breakpoint is created automatically, which will detect when the property we have specified changes, as shown in *Figure 5.26*:

Figure 5.26 – The data breakpoint being created in the Breakpoints window

Let's say, subsequently, we proceed to modify the property data – for example, through the following code:

```
app.Environment.EnvironmentName = "Testing Data Breakpoints";
```

Now, when we run the application, we will see how the breakpoint previously placed is reached, as shown here:

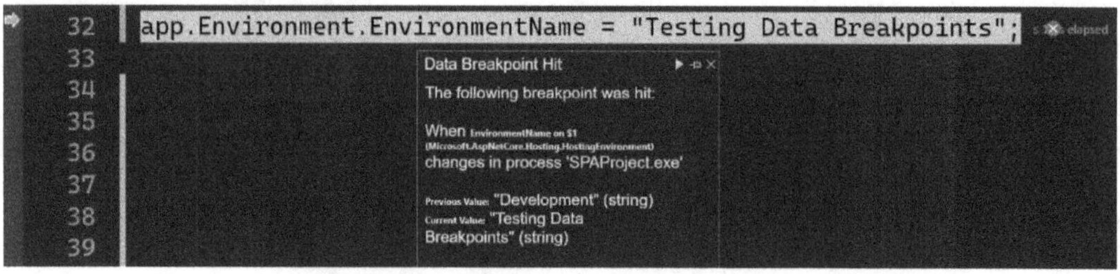

Figure 5.27 – A demonstration of a break when the property has been changed

Finally, you will see that when you restart the application or stop it, the breakpoint will disappear from the window. This is because the object reference is lost.

Dependent breakpoints

This type of breakpoint is a special breakpoint that will only be executed if another breakpoint is reached first. Perhaps in a simple scenario, it is not very useful, but in complex scenarios, where hundreds of functionalities come into play, it can be of great help.

For example, imagine that you have a method that is invoked in several places in your application, and you are testing a new functionality that invokes it. If you were to place a normal breakpoint on the method, it would stop every time it is invoked. With a dependent breakpoint, you can specify that you only want to stop execution if the breakpoint of your new functionality is reached.

To demonstrate this functionality, I have added a couple of methods to the `Program.cs` file, as follows:

```
void NewMethod()
{
    Console.WriteLine("New Method");
}


void CommonMethod(string message)
{
    Console.WriteLine(message);
}
```

The idea of the CommonMethod function is that we can see the content of a string passed as a parameter and know at what point the breakpoint has stopped. With this in mind, let's add some additional lines in which we will first call CommonMethod, then NewMethod, and finally, CommonMethod again:

```
CommonMethod("Before invocation of NewMethod()");
NewMethod();
CommonMethod("After invocation of NewMethod()");
```

To place a dependent breakpoint, we will first place a normal breakpoint in the line of code on which we want to depend – that is, the one that first must be executed for the dependent breakpoint to be executed. In our example, we will place it in the NewMethod functionality, since it is the method we want to test, as shown in *Figure 5.28*:

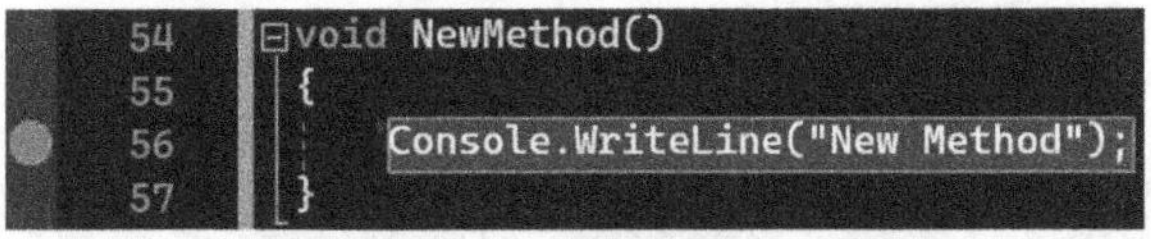

Figure 5.28 – The placement of the breakpoint on which a dependent breakpoint will depend

Then, you must right-click on the line where you want to create the dependent breakpoint, as shown in *Figure 5.29*:

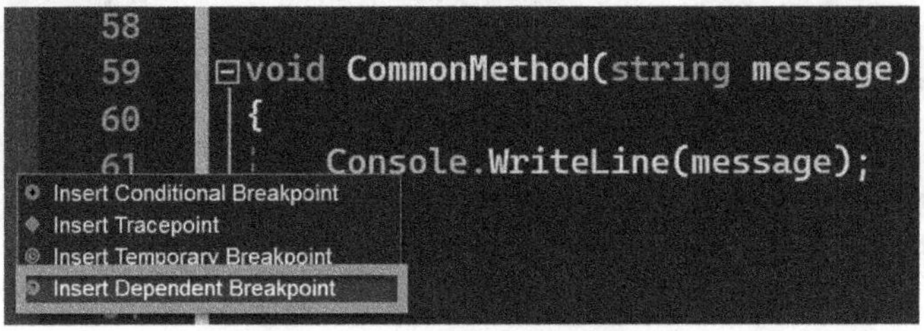

Figure 5.29 – Inserting a dependent breakpoint

This option will display the **Breakpoint Settings** window, where you will be asked which breakpoint you want to depend on to launch the dependent breakpoint. In our example, we will select the only breakpoint that is part of our project, as shown here:

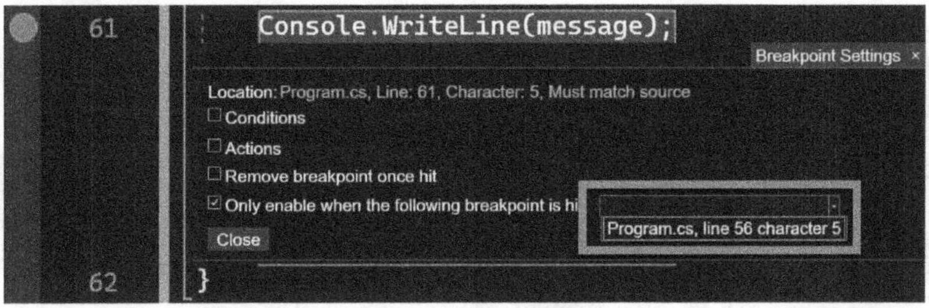

Figure 5.30 – A list of breakpoints on which we can depend for dependent breakpoint execution

Once this window is closed, you will see how a special breakpoint is created that (if you hover over it) will tell you on which other breakpoint it depends, as shown here:

```
 59      ⊟void CommonMethod(string message)
 60       {
🔴 61          ¦   Console WriteLine(message);
Location: Program.cs, line 61 character 5

Dependent on: Program.cs, line 56 character 5
```

Figure 5.31 – A view of a dependent breakpoint

Finally, when running the application, you will see how it stops first on the NewMethod call and not on CommonMethod, even though it has been invoked first. If you continue the execution, the dependent breakpoint will stop, showing the **After invocation of NewMethod()** message, as shown in *Figure 5.32*:

```
                                           📄 message    🔍 View ▾  "After invocation of NewMethod()" ◢□
 54      ⊟void NewMethod()
 55       {
 56          Console.WriteLine("New Method");
 57       }
 58
 59      ⊟void CommonMethod(string message) ▸
 60       {                              📄 message  🔍 View ▾  "After invocation of NewMethod()"
🔴61🖊        Console.WriteLine(message);  ≤ 1ms elapsed
 62       }
```

Figure 5.32 – The execution of a dependent breakpoint

As you can see, this type of breakpoint can be very helpful to avoid constant method invocation.

Temporary breakpoints

Temporary breakpoints, as their name indicates, are breakpoints that are automatically deleted once they are executed. To place one of these, you must right-click on the breakpoint line and select the **Insert Temporary Breakpoint** option, as shown in *Figure 5.33*:

Figure 5.33 – The menu to insert a temporary breakpoint

If you run the application, you will see that once the application has stopped at the breakpoint, it will be automatically deleted. This type of breakpoint can be used when we want to evaluate, for example, the first iteration of a cycle.

> **Important Note**
> To insert a temporary breakpoint quickly, use the *F9 + Shift + Alt* keys, followed by the *T* key.

Now that we have examined the different types of breakpoints available in Visual Studio, let's see how we can take advantage of them using different inspection tools, which we will see in the next section.

Inspection tools for debugging

When working with breakpoints, it is important that we know where to find the information we want to visualize and to know whether it is correct or not. For this, within Visual Studio, we have a series of windows that will allow us to visualize different types of information. So, let's look at them.

Watch window

The **Watch** window will allow us to keep track of the values of variables or properties while we execute our code step by step. It is especially useful when we have pieces of code that are repeated several times, such as cycles or common methods. To access this window, we must first place a breakpoint in the code and execute the application.

Once the application stops at the breakpoint, we will be able to deploy the **Debug | Windows** menu. This will show us a set of new debugging windows that we can only access while running the application. Let's select the **Watch** option to choose a window, as shown in *Figure 5.34*:

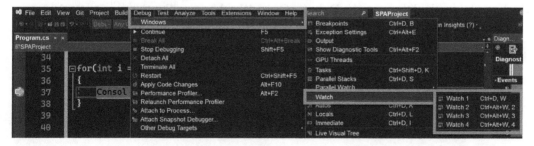

Figure 5.34 – The menu to reach the Watch windows

Once the window has been opened, we can add the name of different variables and properties that we want to monitor. As long as we have entered a valid variable name for the scope we are in, we will be shown its corresponding value in the **Value** column, as shown in *Figure 5.35*:

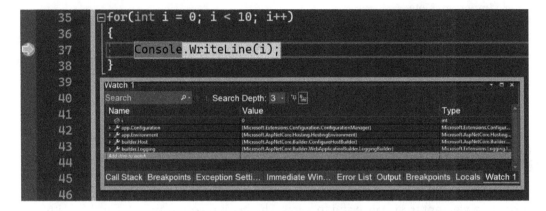

Figure 5.35 – The monitoring of variable values and properties through the Watch window

Another very simple way to add variables to a **Watch** window is to right-click on the variable you want to monitor when the application is running and select the **Add Watch** option.

> **Important Note**
>
> To access any **Watch** window quickly, you can use the *Ctrl + Alt + W* shortcut, followed by the window number (from *1* to *4*).

QuickWatch

The **Watch** window is very useful for tracing values through the application. But if we want to use the same functionality as the **Watch** Window, to test individual expressions without keeping the same ones, then we can choose to use the **QuickWatch** window, which can be reached through the **Debug | QuickWatch** menu. For the demonstration, we will use the example of the for loop (seen in the *Hit Count* section), as shown here:

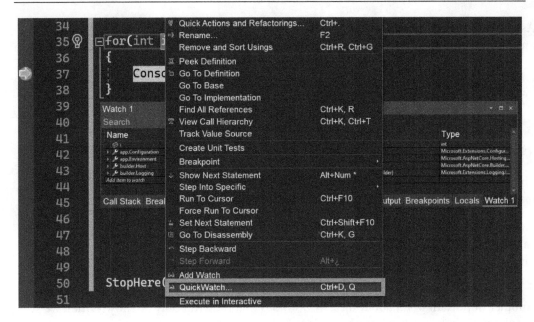

Figure 5.36 – The QuickWatch option from the Debug menu

This option will activate a modal window, which means that we will not be able to continue running the application until we close it. In this window, we will be able to see data such as the name of the expression, the value, and the type of data we are evaluating.

Something very useful in this window is that we will be able to modify the expression. This means that we will be able to evaluate other variables and can even execute some operations that we want to test. In *Figure 5.37*, we see this in action, having entered a new expression to evaluate whether its result is `true` or `false`:

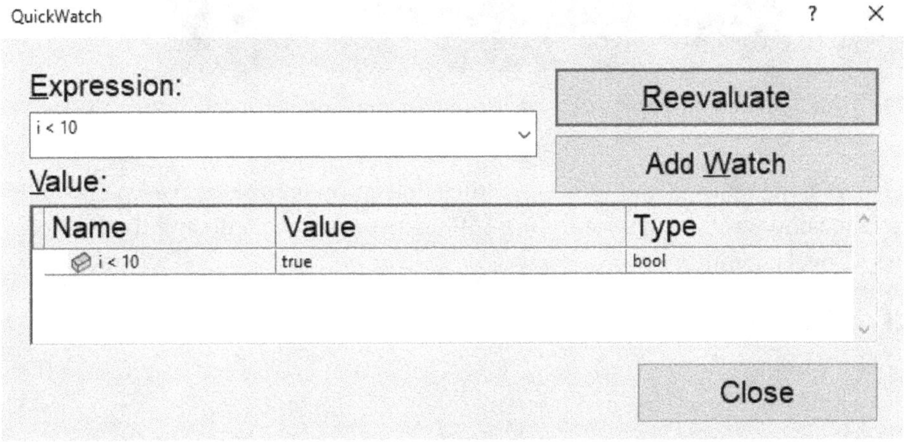

Figure 5.37 – Modifying expressions in the QuickWatch window

As we can see, this offers a huge advantage, since we can test variables and properties that are quickly available in real time.

> **Important Note**
> You can use the *Shift + F9* shortcut to access the **QuickWatch** window quickly.

The Autos and Locals windows

The **Autos** and **Locals** windows allow us to view information about variables and properties without the need to add them somewhere as with the **Watch** window.

However, these have a specific scope. While the **Autos** window shows the value of the variables around the breakpoint we have placed, the **Locals** window will show values only for the current scope – that is, usually the function or method in which the breakpoint is located.

Another important point about these windows is that they will only be shown when we run the application and after we have placed a breakpoint.

For the demonstration of this window pair, I have added a new method to our `Program.cs` file and placed a breakpoint, as shown in *Figure 5.38*:

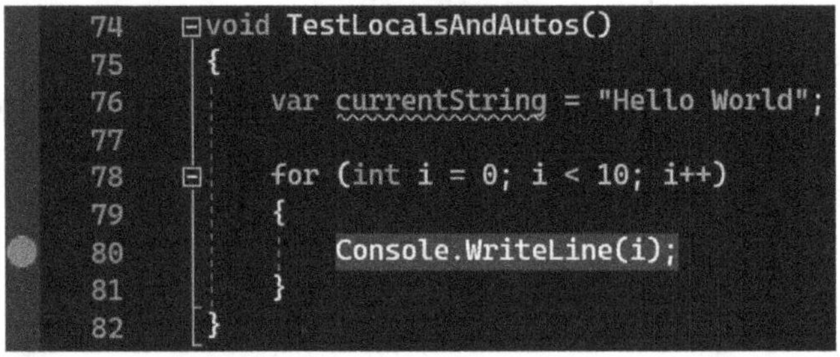

```
74    void TestLocalsAndAutos()
75    {
76        var currentString = "Hello World";
77
78        for (int i = 0; i < 10; i++)
79        {
80            Console.WriteLine(i);
81        }
82    }
```

Figure 5.38 – The TestLocalsAndAutos method, with a breakpoint placed to demonstrate the Locals and Autos windows

Once we invoke the method and only when the application is running, we will be able to display the windows through the **Debug | Windows | Autos** menu and the **Debug | Windows | Locals** menu:

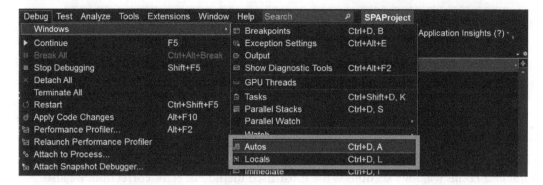

Figure 5.39 – Accessing the Autos and Locals windows from the Debug menu

Let's first examine the **Locals** window. We will see how it contains all the local variables that are available, including the `i` and `currentString` variables, which belong to the `TestLocalsAndAutos` method:

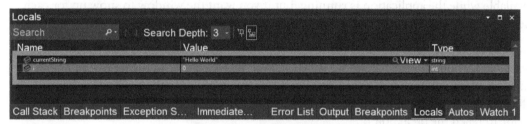

Figure 5.40 – The Locals window

On the other hand, if we look at the **Autos** window, we will see only the content of the `i` variable, since it is the information that is in the scope of the breakpoint, which means the `for` loop:

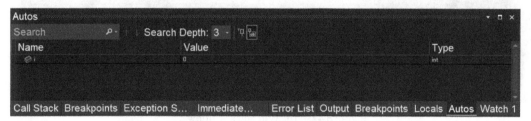

Figure 5.41 – The Autos window

This pair of windows is tremendously useful to see all the available information on our variables at a glance.

> **Important Note**
>
> It is possible to display these windows through the following shortcuts:
>
> **Autos**: *Ctrl + D*, followed by the *A* key
>
> **Locals**: *Ctrl + D*, followed by the *L* key

Call stacks

Although the use of windows that we saw in the *Watch window, QuickWatch*, and *The Autos and Locals windows* sections is the most-used method of error correction, we also have some others to keep track of the invocation flow between our methods.

One of such window is called **Call Stack**, which when we place a breakpoint, will show the stack of calls that have been made until arriving at the same one. To show the window, we must select the option from **Debug | Windows | Call Stack**. This menu item will only be available while the application is running, but if you want the window to show up for each debug session, you can use the thumbtack/pin icon to dock it.

In *Figure 5.42*, we have executed the same code as that in the *The Autos and Locals windows* section, stopping at the same breakpoint. As we can see in the following figure, the **Call Stack** window shows the line of code where we are once the application has stopped due to the breakpoint, as well as the set of calls between methods that have been made to reach the breakpoint:

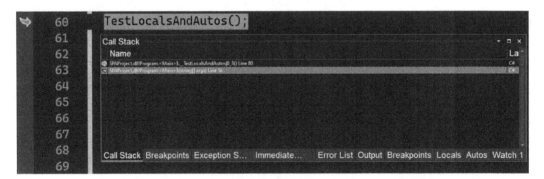

Figure 5.42 – The Call Stack window

In *Figure 5.42*, we can also see the state of the application in each one of the calls if we double-click on any of them. This will allow us to see the values of properties and variables at some specific point in the flow of the application.

This will take us to the code section from where we invoke the method, and it will be highlighted with a different color so that we don't get confused with the colors of the breakpoint.

> **Important Note**
> To display the **Call Stack** window, you can also press the *Ctrl + Alt + C* keys.

Immediate window

Another window of great interest is the **Immediate** window. This window, which we can open from the **Debug | Windows | Immediate** menu, will allow us to make evaluations of expressions, execute statements, and print values of variables and properties.

When you open this window, you will surely be surprised to see nothing inside it. This is because this window will allow you to write on it, to enter the name of a variable or property, and even invoke methods (if they are within the scope), added to the fact that you will have available all the potential of IntelliSense while you write your expressions. IntelliSense is Visual Studio's auto-completion help enhanced with artificial intelligence, and we will talk more about it in *Chapter 7, Coding Efficiently with AI and Code Views*.

In *Figure 5.43*, the invocation of the `TestLocalsAndAutos` method has been executed again, and we have obtained a substring of the main `currentString` string:

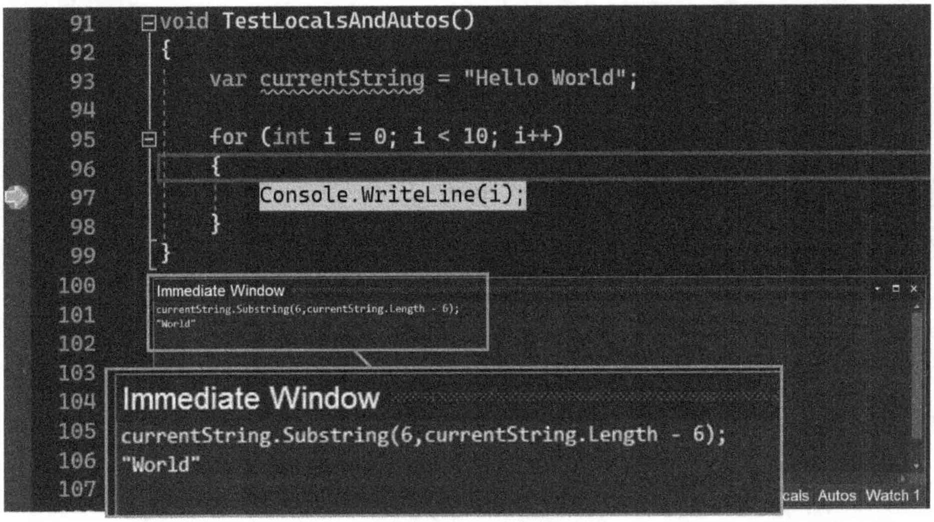

Figure 5.43 – The Immediate Window

A great advantage of this window is that we can execute as many expressions as we want, keeping all the list of results we obtain, as well as the expressions entered.

> **Important Note**
> You can display the **Immediate** window if you press the *Ctrl + Alt + I* keys.

Summary

Visual Studio has many options for debugging source code. In this chapter, we have learned what breakpoints are, the different types of breakpoints, and the associated windows that we can activate to keep track of data in variables and properties. This information has been of utmost importance, since it will help you to solve problems in your code when you face them.

In the next chapter, *Chapter 6, Adding Code Snippets*, you will learn about the concept of code snippets, which allow you to reuse common pieces of code across different projects, quickly adjusting them to your needs.

Part 2: Tools and Productivity

In this part, you will learn how to take advantage of the tools provided by Visual Studio 2022 and how to improve your productivity with some tips and hacks.

This part contains the following chapters:

6
Adding Code Snippets

Coding is a lovely activity, but at times, we have some repeated statements to solve a few known situations, which makes coding more of a monotonous process. **Code snippets** are a good resource to reuse pieces of code where it's desirable. Visual Studio has some code snippets by default that we can use while we are coding, but we also have some tools to create our own code snippets.

In this chapter, we will learn about how Visual Studio helps us to write code faster using code snippets and how to create our own.

We will review the following topics and functionalities for snippets:

- What are code snippets?
- Creating code snippets
- Deleting code snippets
- Importing code snippets

Let's start recognizing the concept of code snippets and how they work in Visual Studio.

Technical requirements

To complete the demos of this book chapter, you must have previously installed Visual Studio 2022 with the web development workload, as shown in *Chapter 1, Getting Started with Visual Studio 2022*. It's important to have the SPA base project that we created in *Chapter 4, Creating Projects and Templates*.

You can check the changes made in `WeatherForecastController.cs` at the following link: `https://github.com/PacktPublishing/Hands-On-Visual-Studio-2022/blob/main/Chapter06/WeatherForecast.cs`.

What are code snippets?

Code snippets are a simple and easy way to reuse code by creating templates that generate common statements, such as conditionals, loops, or comment structures.

Visual Studio has many code snippets by default for almost all the supported technologies and programming languages. There are many ways to use code snippets in Visual Studio, so let's check them out.

> **Important Note**
>
> Code snippets are a common concept in software development. Almost all IDEs and code editors provide code snippets or have extensions to include code snippets.

Using the SPA project created in *Chapter 4, Creating Projects and Templates*, you will create a new condition to return an empty collection in the `Get` method by navigating to the `WeatherForecastController.cs` file. Just write the word `if` to see the code snippet suggested by Visual Studio (see *Figure 6.1*):

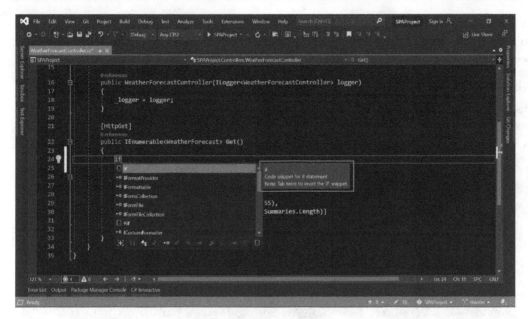

Figure 6.1 – The code snippet for the if statement suggested by Visual Studio

Since a conditional statement is a very common piece of code, Visual Studio gives you the option to create this code quickly. You can click on if or continue writing if you don't want to perform any action. You can also press *tab* twice to create the if statement automatically.

There is an option in the intelligent code completion (also called **IntelliSense**, about which we will talk more in *Chapter 7, Coding Efficiently with AI and Code Views*) suggestions where you can see all the code snippets filtered by the characters that you wrote. See the code snippets filter marked in red in *Figure 6.2*:

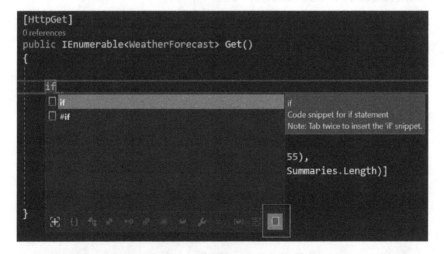

Figure 6.2 – The code snippets filter in Visual Studio

When the code snippet for the `if` condition is highlighted, you can press the *tab* key twice to generate the code of the `if` statement, including the brackets (*see Figure 6.3*):

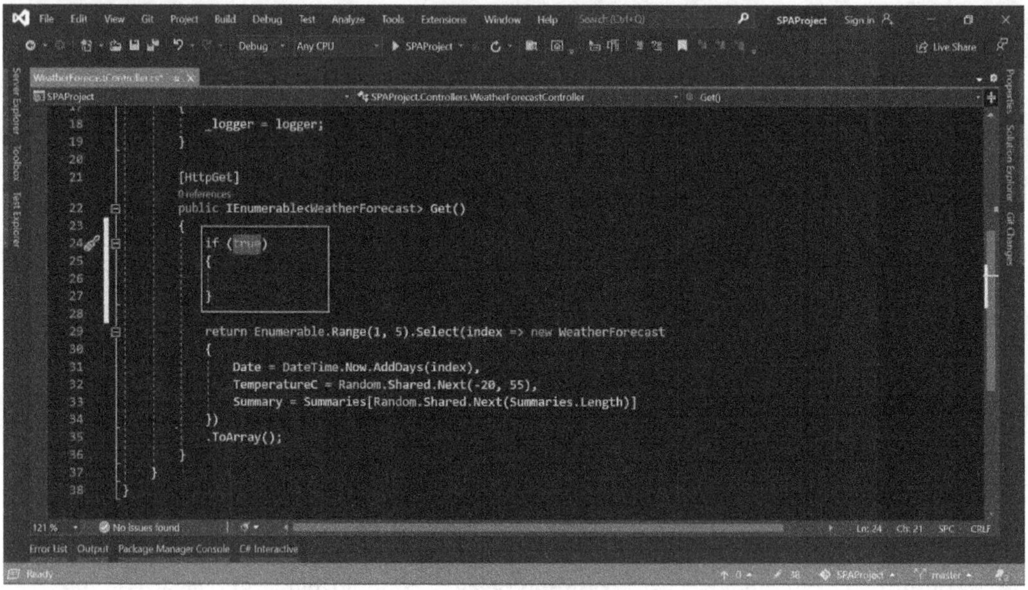

Figure 6.3 – The if statement created by Visual Studio

You will get the `if` statement, including the braces and `true` as a default value. You need to replace `true` with your condition. In this case, you can add a condition to return an empty collection when the operating system is Linux:

```
[HttpGet]
public IEnumerable<WeatherForecast> Get()
{
        if (OperatingSystem.IsLinux())
        {
            return new List<WeatherForecast>();
        }

        return Enumerable.Range(1, 5).Select(index =>
            new WeatherForecast
        {
            Date = DateTime.Now.AddDays(index),
            TemperatureC = Random.Shared.Next(-20, 55),
            Summary = Summaries[Random.Shared.Next
                (Summaries.Length)]
```

```
        })
        .ToArray();
}
```

In the previous code block, we added a condition in the `Get` method before the default logic to check whether the operating system where the app is running is Linux or not. Within the condition, we are returning an empty list.

There are many useful code snippets included as default for C#, but the following are the most popular:

- `try`: Creates a structure for a `try/catch` statement
- `for`: Generates a `for` statement using the local `i` variable
- `ctor`: Creates the constructor of the class automatically
- `switch`: Generates a `switch` statement
- `prop`: Creates a new property in the current class

You can try some of these code snippets in C# code to see the code generated by Visual Studio and use them when the need arises.

Let's see another example using a CSS file. Navigate to **ClientApp | src** and click on the `custom.css` CSS file (*see Figure 6.4*):

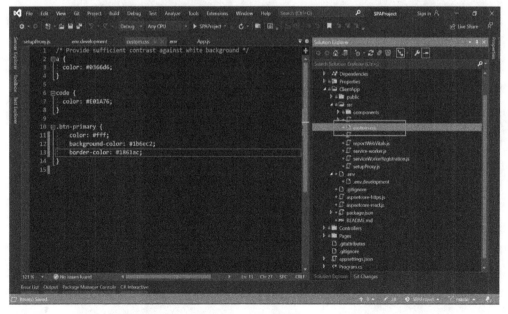

Figure 6.4 – The custom.css file loaded in Visual Studio

In the `custom.css` file, you can write the `columns` property and see how Visual Studio suggests a code snippet for it. See the code snippets suggested for this demo in *Figure 6.5*:

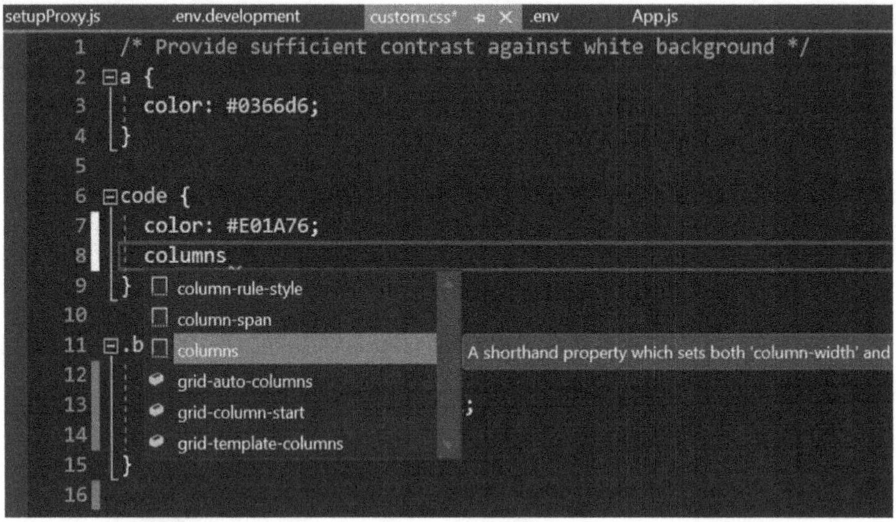

Figure 6.5 – The code snippets suggested by Visual Studio when you type the word "columns"

Once again, using the key *tab* twice, you can generate the code for this property automatically. See the code generated in *Figure 6.6*:

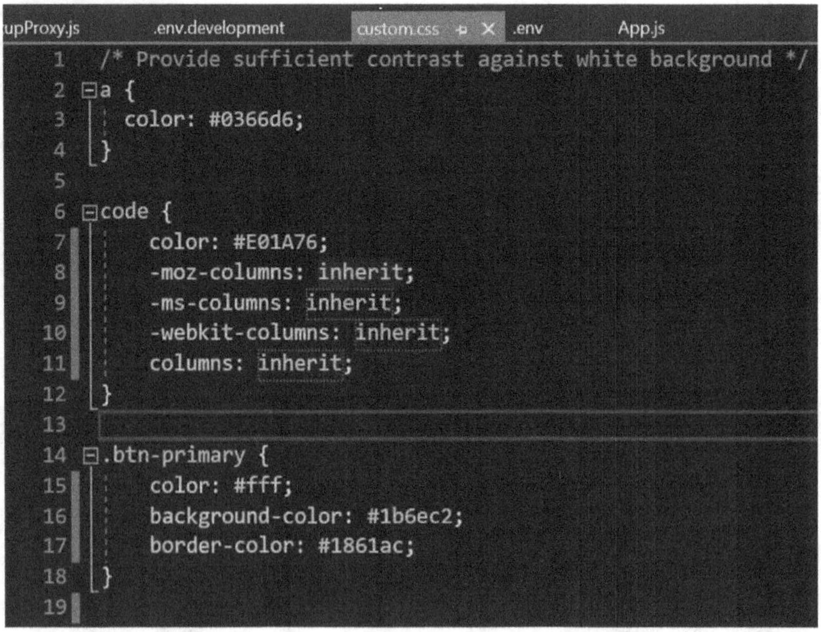

Figure 6.6 – The code generated by the code snippet

In this case, Visual Studio is going to generate four properties, one per browser, to ensure the code is compatible with all of them (Chrome, Mozilla, and others). We can keep the default `inherit` value because the proposal of this action is just to see how Visual Studio makes the code.

You are now ready to use code snippets in Visual Studio. You can identify which suggestions by Visual Studio are code snippets and how to filter them. Now, it's time to learn how to create your own code snippets and use them in your code.

Creating code snippets

To create a code snippet in Visual Studio, we need to create a file with the `snippet` extension. This file has an XML format, and there is a base template that we can update to include the information for our code snippet. The following code is a template example:

```
<?xml version="1.0" encoding="utf-8"?>
<CodeSnippets xmlns="http://schemas.microsoft.com/
    VisualStudio/2005/CodeSnippet">
    <CodeSnippet Format="1.0.0">
        <Header>
            <Title></Title>
            <Author></Author>
            <Description></Description>
            <Shortcut></Shortcut>
        </Header>
        <Snippet>
            <Code Language="">
                <![CDATA[]]>
            </Code>
        </Snippet>
    </CodeSnippet>
</CodeSnippets>
```

Let's review all the properties in this XML and understand how to create our first code snippet.

In the `Header` section, we have the following:

- `Title`: Name or general information
- `Author`: Creator or author

- `Description`: What your code snippets do

- `Shortcut`: The shortcut to call the code snippet when you are typing

In `Snippet`, we have the following:

- `Language`: The programming language for the code

- `[CDATA[]]`: Contains the code

Now, we can create a code snippet to detect whether the operating system where the code is running is Linux or not:

```xml
<?xml version="1.0" encoding="utf-8"?>
<CodeSnippets xmlns="http://schemas.microsoft.com/
VisualStudio/2005/CodeSnippet">
    <CodeSnippet Format="1.0.0">
        <Header>
            <Title>If Linux condition</Title>
            <Author>Myself</Author>
            <Description>Conditional to now if the
                operating system is Linux</Description>
            <Shortcut>ifln</Shortcut>
        </Header>
        <Snippet>
            <Code Language="CSharp">
                <![CDATA[if (OperatingSystem.IsLinux())
                {
                    return new List<WeatherForecast>();
                }]]>
            </Code>
        </Snippet>
    </CodeSnippet>
</CodeSnippets>
```

You can create a new folder in documents or any new folder for this activity and save the file with the `snippet` extension (see *Figure 6.7*):

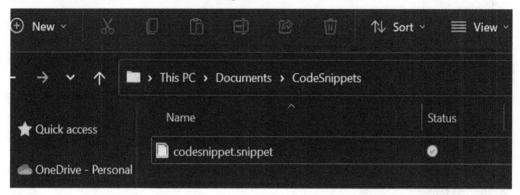

Figure 6.7 – The code snippet file in Windows explorer

Now, the last step is to add this folder to the code snippet section in Visual Studio. Navigate to **Tools | Code Snippets Manager**, and in the **Language** dropdown, select **CSharp** (see *Figure 6.8*):

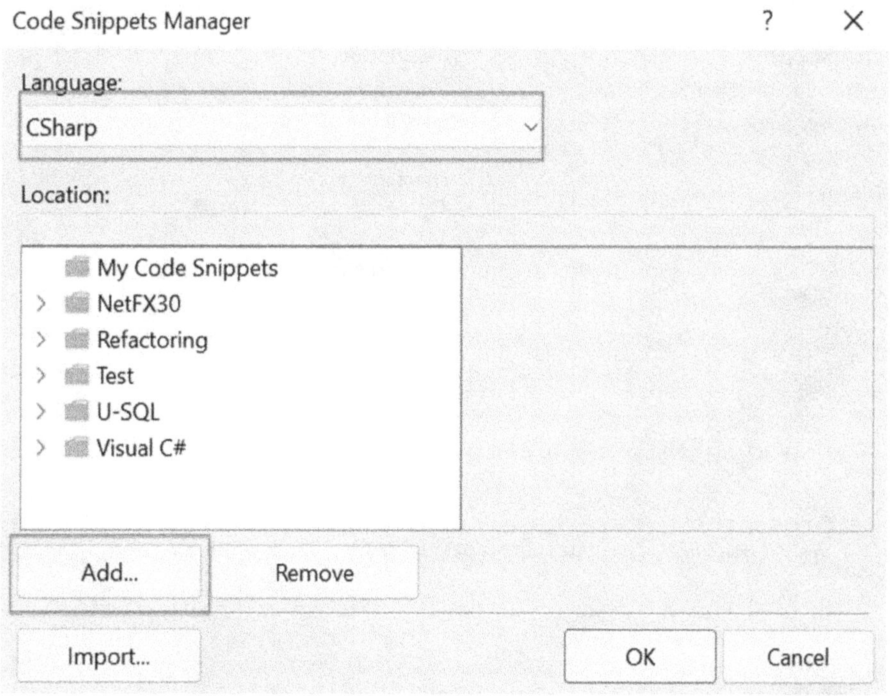

Figure 6.8 – Code Snippets Manager in Visual Studio

> **Important Note**
> You can use the *Ctrl + K* shortcuts followed by *Ctrl + B* to open the **Code Snippets Manager**.

The **Language** option includes all the programming languages and technologies supported by Visual Studio, depending on the workload installed.

You can click on **Add...** and select the folder where your code snippet was created.

After adding the folder, you will see a new folder in the list, including the new code snippet. If you select this code snippet, you will see the details on the right panel, as shown in *Figure 6.9*:

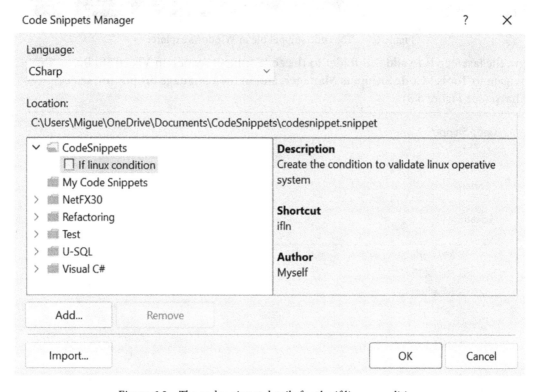

Figure 6.9 – The code snippet details for the if linux condition

Now, you are ready to use your code snippet in any C# file. Open the
`WeatherForecastController.cs` file and try your new code snippet in the
`Get` method:

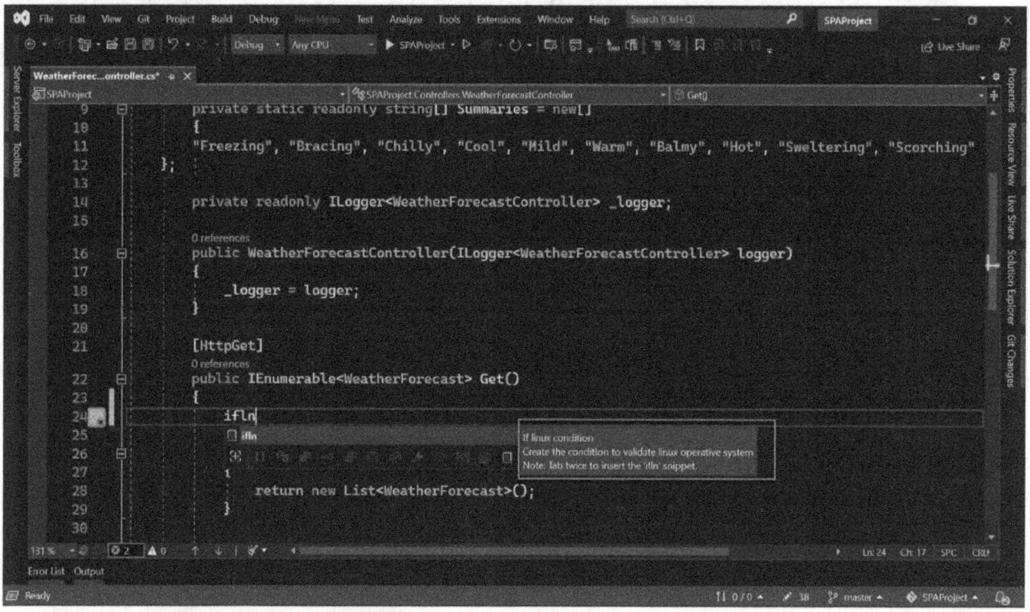

Figure 6.10 – Using the if linux condition code snippet in Visual Studio

You can see in *Figure 6.10* how Visual Studio suggests your new code snippet, and in the
tooltip, there is the description that you provided. As usual, you can press *tab* twice to
generate the code for the code snippet.

Important Note

For more information about how to create and design code snippets, look
at the official documentation: `https://docs.microsoft.com/`
`visualstudio/ide/code-snippets`.

So far, you have created your first code snippet and know how to create others that meet
your needs. But there are also other actions that you can perform with code snippets. So,
let's see how to delete a code snippet in the next section.

Deleting code snippets

Due to human error, we can add code snippets that we don't need, or maybe we selected the wrong code snippet. For these scenarios, Visual Studio has an option to delete code snippets. To see this option, navigate to **Tools | Code Snippets Manager** and select the `CodeSnippets` folder. This folder contains the code snippet that you included in the *Creating code snippets* section. If you used a different name, select the correct folder for you. You can see the **Remove** button location in *Figure 6.11*:

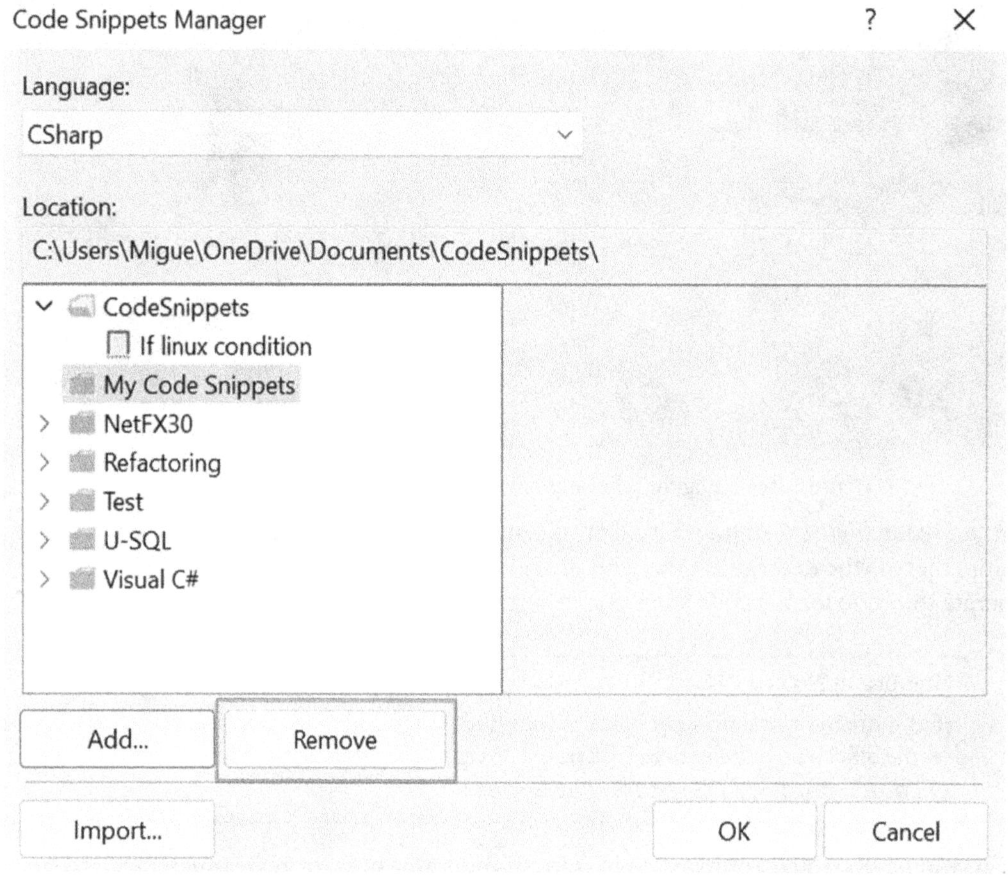

Figure 6.11 – The Remove button in Code Snippets Manager

The **Remove** button will delete the whole folder, including all the code snippets inside. In Visual Studio 2022, it's not possible to remove code snippets one by one, and therefore, we need to create a folder with a proper name for our code snippets. After removing the folder, Visual Studio is not going to suggest the code snippets anymore.

> **Important Note**
> When you remove code snippets in Visual Studio, the original files and folder are not removed from your local system. Only the reference to the file in Visual Studio will be removed.

At this point, you know how to create and delete code snippets. We can also import code snippets in Visual Studio, so let's see how to do it.

Importing code snippets

If we want to include code snippets in a folder already created in **Code Snippets Manager**, we can use the **Import...** option:

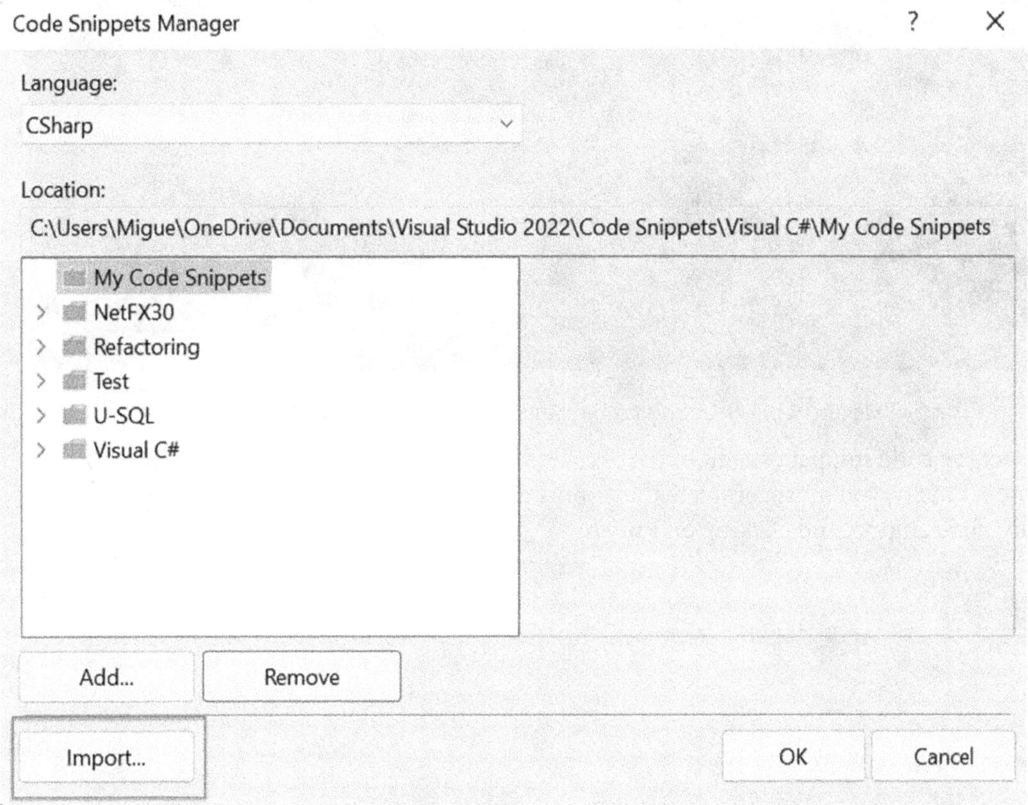

Figure 6.12 – The Import... button in Code Snippets Manager

After clicking on **Import…**, you need to select the code snippet that you want to import in the selected folder. There is a filter related to the `.snippet` extension in the modal (see *Figure 6.13*):

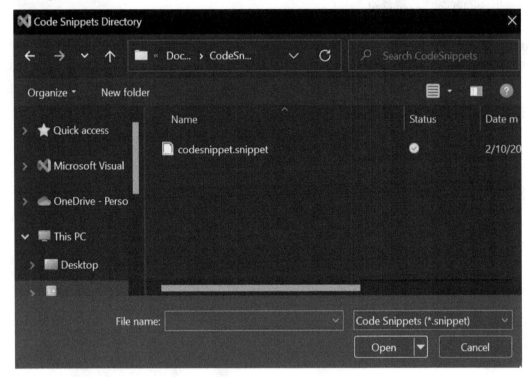

Figure 6.13 – Selecting a code snippet file (.snippet) in the filesystem

Select the code snippet created in the *Creating code snippets* section, and then click on **Open**. Finally, you must complete the import process by selecting the location folder for your code snippet and clicking on **Finish**:

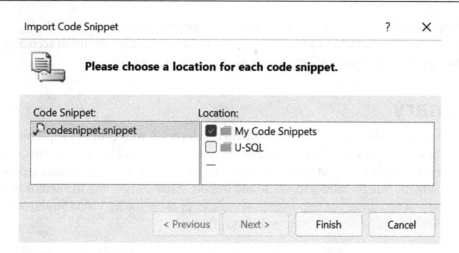

Figure 6.14 – The selection location for the imported code snippet

After completing the importation, you will see that the code snippet is added to the selected folder. See the imported code snippet in *Figure 6.15*:

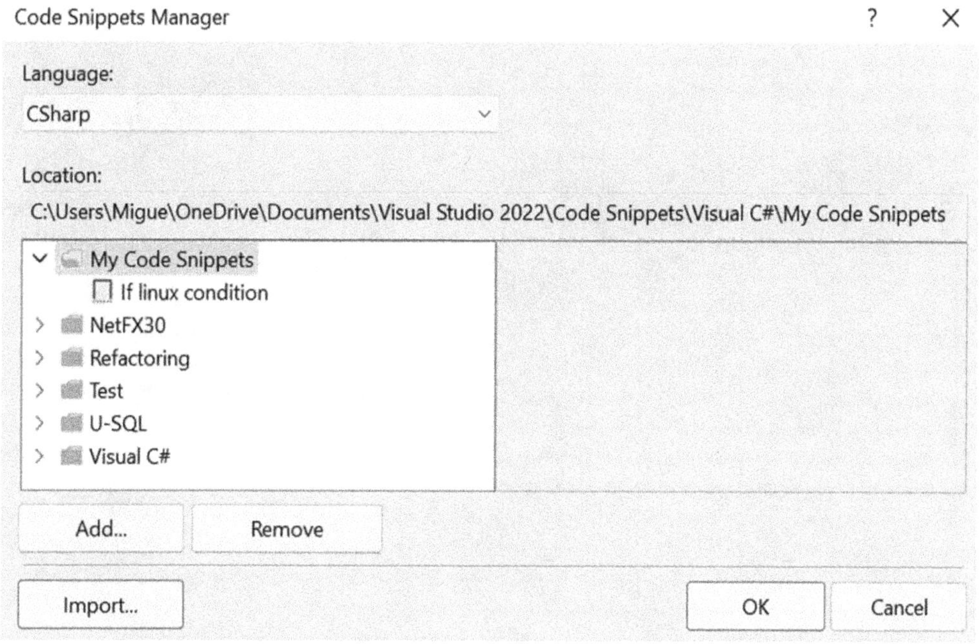

Figure 6.15 – The code snippet imported into the My Code Snippets folder

Importing code snippets is a great way to share our custom snippets with friends, colleagues, and coworkers. We can create amazing pieces of code for common scenarios and some special code closely related to our architecture or guidelines.

Summary

Now, you can use code snippets in Visual Studio and increase your productivity. You can identify which pieces of code are common in your architecture and use patterns to create your own code snippets to meet your requirements. Also, you know how to manage code snippets using the functionalities to delete and import. After completing the demos in this chapter, you will recognize the importance of code snippets and why Visual Studio is a powerful IDE that helps developers to write code faster.

In *Chapter 7, Coding Efficiently with AI and Code Views*, we will review the **artificial intelligence** (**AI**) included in Visual Studio and how this tool can help us to write code faster and improve the syntax in some scenarios. You will also do some demos where the AI will help you, allowing you to predict what action or statement you want to perform.

7
Coding Efficiently with AI and Code Views

Artificial intelligence is a vast and interesting field that allows us to improve our lifestyle in one way or another. We see, hear, and use it every day, and if you don't believe it, ask yourself how many times you use the Google search engine throughout the day. Other places where you can find it are in photo-editing programs, where it is possible to remove, for example, the background of an image in an almost perfect way. Social networks are another perfect example of the use of artificial intelligence, as they are constantly processing the best recommendations for you to stay on them as long as possible.

Fortunately, artificial intelligence has even reached new software development tools, through predictive code integration, which allows us to choose the pieces of code we need at the right time. In Visual Studio, we have a powerful feature called **Visual Studio IntelliCode** that does this.

Similarly, we have different visual tools and windows that can help us find relationships in our code and navigate through them efficiently.

In this chapter, we will cover the following main topics:

- Understanding CodeLens
- Working with code views
- Using Visual Studio IntelliCode

Technical requirements

To use IntelliCode, Visual Studio 2022 with the web development workload must be installed. Because this chapter focuses more on showing auto-completion and class display features, no changes have been made to the code repositories. You can use the repository hosted at the following URL: `https://github.com/PacktPublishing/Hands-On-Visual-Studio-2022/tree/main/Chapter07`.

Also, to be able to perform the procedure of the *Code maps* section, it is required to use the Enterprise version of Visual Studio.

Likewise, the **Code Map** and **Live Dependency Validation** tools must be installed by the Visual Studio Installer, selecting them as shown in *Figure 7.1*:

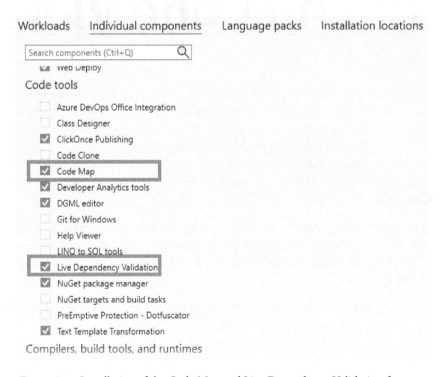

Figure 7.1 – Installation of the Code Map and Live Dependency Validation features

Now that we know the technical requirements, let's learn how to work with them to get the most out of Visual Studio.

Understanding CodeLens

CodeLens is a powerful set of tools that is useful for finding references in code, relationships between your different components, seeing the history of changes in the code, linked bugs, code reviews, unit tests, and so on.

In this section, we will analyze the most important tools of this feature. Let's start by seeing how we can find references in our code.

Finding references in code

CodeLens is presented in our code files from the first time we use Visual Studio. We can check this by going to any class, method, or property and verifying that a sentence appears, indicating the number of references in the project about it. In *Figure 7.2*, we can see that we have opened the `WeatherForecastController.cs` file, which shows us that three references have been found for the `WeatherForecastController` class:

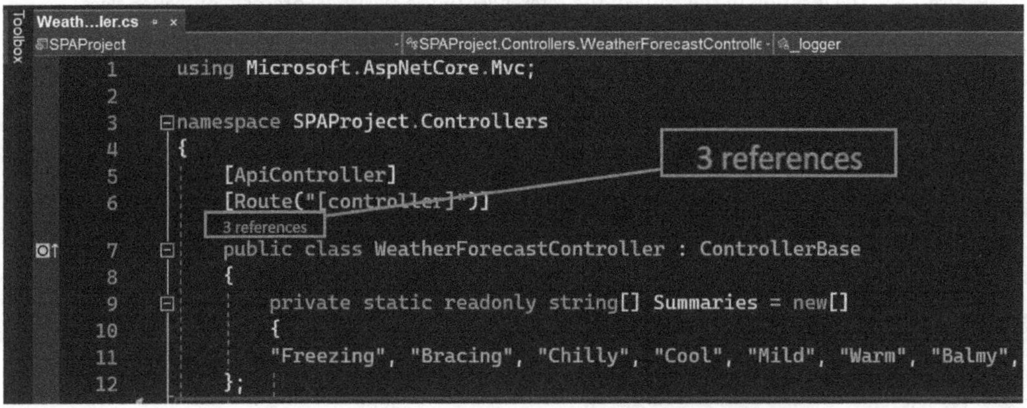

Figure 7.2 – References for the WeatherForecastController class

This means that the `WeatherForecastController` class is being used in three places in our project. If we proceed by clicking on the legend titled **3 references**, as shown in *Figure 7.2*, we will see all references that use this class. In *Figure 7.3*, continuing with the example, we can see that it is used within the same class in which we are – that is, the `WeatherForecastController` class – specifically in the **14** and **16** lines:

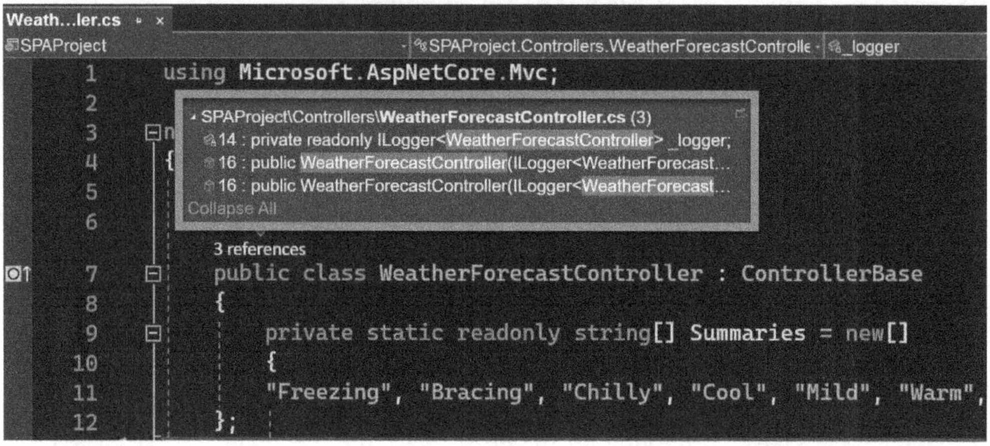

Figure 7.3 – The location in the code of the references in the WeatherForecastController class

Not only that, but we can also position over on any of the lines found, which will show us a section of the four closest lines of code surrounding the reference. This way, we can get a better idea of the purpose of using it, as shown in *Figure 7.4*:

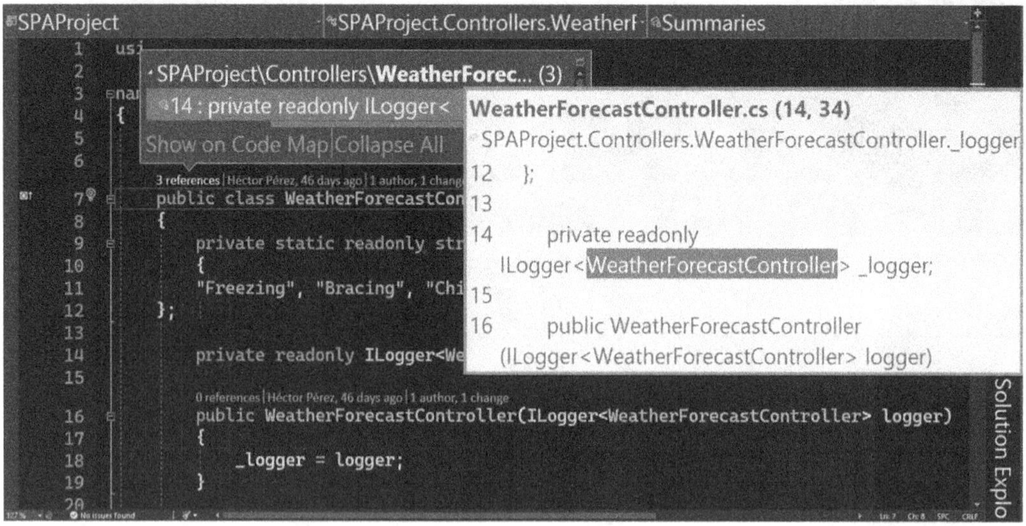

Figure 7.4 – A preview of a found code reference

This is quite useful if we are in a new project and need to quickly know what certain parts of the code do.

Now, let's see a utility belonging to CodeLens that will allow you to see relationships between code visually.

> **Important Note**
>
> Sometimes, even if the number of references equals zero, there may be references to other GUI files, such as `.xaml` and `.aspx` files.

Code maps

Code maps are a way to visualize relationships in code in a fast and efficient way. This tool allows the creation, as its name indicates, of visual maps from the code. With this tool, we will be able to see the structure of the entities, their different properties, and relationships, which lets us know how much impact a change we make can have.

There are several ways to create code maps. The first one is by selecting the option **Architecture | New Code Map**. This will open a new document with a `.dgml` extension, in which we will be instructed to drag files from the solution explorer, class view, or object browser, as shown in *Figure 7.5*:

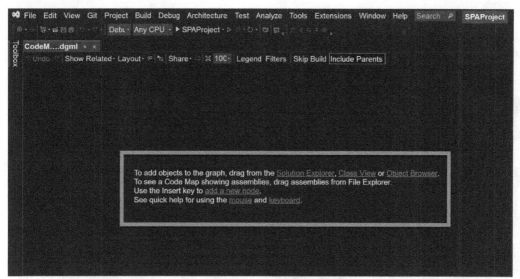

Figure 7.5 – The empty code map file

Let's do a test – click on the **Class View** link to open the class window and then expand the SPAProject.Controllers namespace. This will show you the WeatherForecastController class, as shown in *Figure 7.6*:

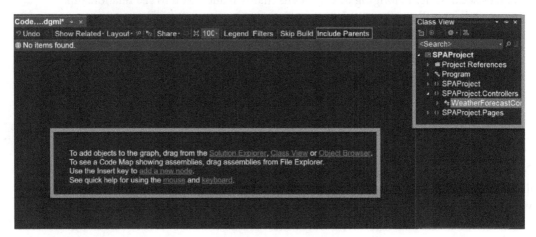

Figure 7.6 – A visualization of the WeatherForecastController class, about to be dragged into the code map file

Next, drag the WeatherForecastController class into the code map file. This will automatically generate a graph where we can see the dragged class, namespace that contains it, and finally, .dll in which it is hosted, as shown in *Figure 7.7*:

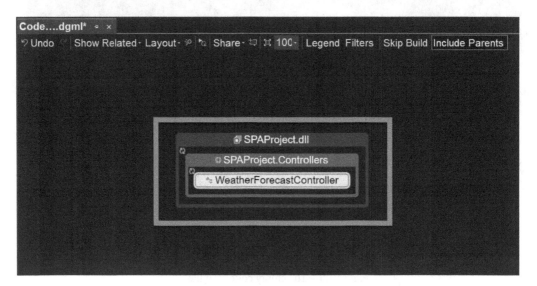

Figure 7.7 – The WeatherForecastController class in the code map

Additionally, if we expand the `WeatherForecastController` class in the diagram, we will see the members that are part of the class, such as its attributes and behavior, as well as the relationships that can be found as part of the same class:

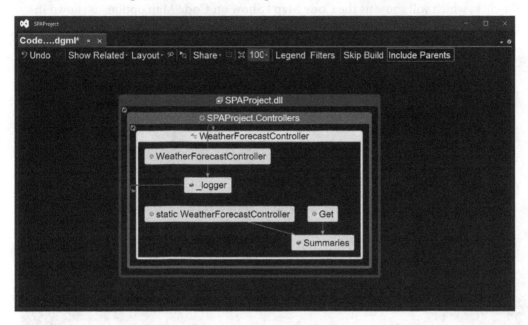

Figure 7.8 – The relationships found through the code map

In *Figure 7.8*, we can see in action a code map of the `WeatherForecastController` class with all its members expanded. This shows us quickly how the fields, properties, and methods are related.

Another way to create a map code from the source code is to go to the file where the member we are interested in is located, such as the `WeatherForecast.cs` file. Once we have opened the file, we can position the cursor on a class, method, property, or field and right-click, which will show us the **Code Map | Show on Code Map** option, as shown in *Figure 7.9*:

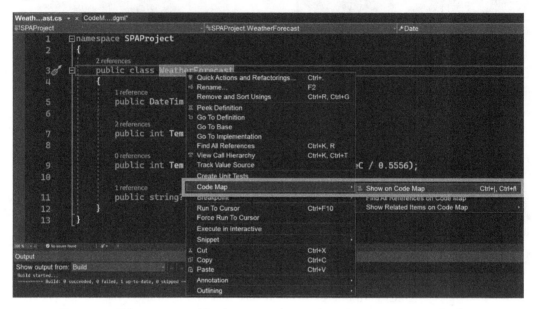

Figure 7.9 – The option to add a class to a code map from the context menu of a class

This option will create a new `.dgml` file or, if you have already created one, as in our case, add the reference with its respective relations in the previously opened file.

> **Important Note**
>
> If you want to center the code map diagram, at any time you can click on any empty area of the diagram to center it. Likewise, if you double-click on any of the entities or members of the diagram, the corresponding code will open to view it next to the diagram.

As a result of adding the new class to the diagram, we can see that the `WeatherForecast` class is being used in the `Get` method of the `WeatherForecastController` class, as shown in *Figure 7.10*. This way, we have discovered this relationship very easily:

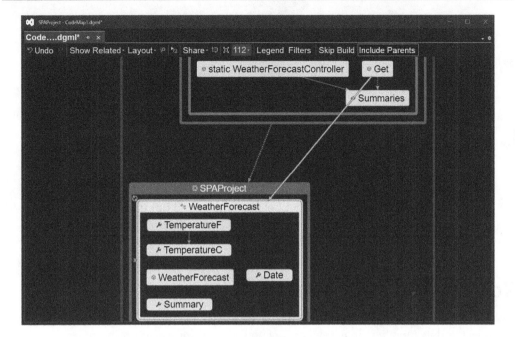

Figure 7.10 – The relationships between classes by code map

> **Important Note**
>
> The arrows indicating relationships between entities in the code map diagram
> appear and disappear as entities are selected, allowing more space and
> understanding of the diagram. I encourage you to select each of the elements in
> the diagram so that you can see the complete relationship.

Finally, if we want to be able to see the relationships in our solution without having to add
entity by entity, from the **Architecture** menu, we can select the **Generate Code Map for
Solution** option, as we can see in *Figure 7.11*:

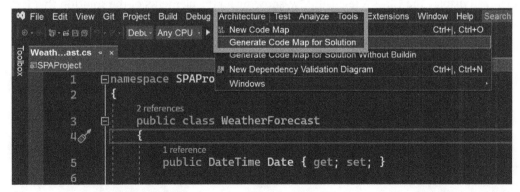

Figure 7.11 – The option in the menu to generate a code map at a solution level

This will start the process of generating the respective code map for the entire solution. Depending on the number of references in your code, the process may take more or less time.

> **Important Note**
> Although code maps can only be created in Visual Studio Enterprise Edition, it is possible to view them from any version of Visual Studio, including the Community version, but it is not possible to edit them from any version except the Enterprise Edition.

Now that we have seen how CodeLens can help us understand our code better and faster, let's look at the windows available in Visual Studio, which will allow us to work with our code easily.

Working with code views

In addition to CodeLens, there are several windows that can help us to examine the classes of a project and its members in a quicker way. In this section, you are going to learn about them and how they can help you in breaking down code of a project in Visual Studio.

Class view

The class view is a window that allows you to see the elements of a Visual Studio project, such as namespaces, types, interfaces, enumerations, and classes, allowing you to access each of these elements quickly. Perhaps if you have worked with small projects in Visual Studio, you might not see it as being of much use. But if, like me, you work with solutions that can have up to 20 projects or more, then it is an excellent option to examine code.

To access this window, you must select the **View | Class View** option from the menu, which will display the **Class View** window, showing all the elements of the solution that is currently open, as shown in *Figure 7.12*:

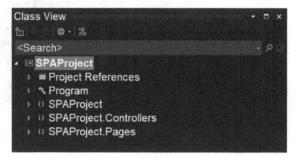

Figure 7.12 – The Class View window with a loaded project

As you can see, we can get a very quick idea of the structure of our project by seeing at a glance the namespaces into which the project has been divided. If we expand the nodes of each of the namespaces, we can also see the different classes that are part of these namespaces, as shown in *Figure 7.13*:

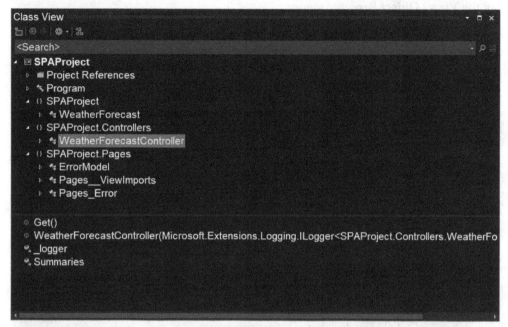

Figure 7.13 – The Class View window displaying the members of the WeatherForecastController class

In addition, if we select any element of our project, such as a class, we will be able to see the properties and methods that compose it in the lower part of the window (as shown in *Figure 7.13*).

Another great advantage of this window is that you don't even have to recompile the project to see the changes, as they will be made automatically and instantly.

At the top of the window in *Figure 7.13*, we can also see a series of buttons that we can use to create new folders, navigate between the selected elements, configure the display options, and possibly add a class to a code map file.

It is certainly an excellent window to navigate between the classes of our project, but what if we want to navigate between classes that are not part of our project? In this case, the object browser can help us, which we will examine next.

The object browser

The object browser is a very useful window that has been present since the beginning of Visual Studio. This window contains information about all the assemblies that are used in your project and allows you to examine them in depth. To access this window, we can do it from the **View | Object Browser** menu.

Once we select the **Object Browser** option, it will open and load the assemblies that are used as part of our solution. We can see that the list of assemblies is quite long, and this is because we can examine assemblies that are part of the framework we are using and also the assemblies that we have created ourselves, as shown in *Figure 7.14*:

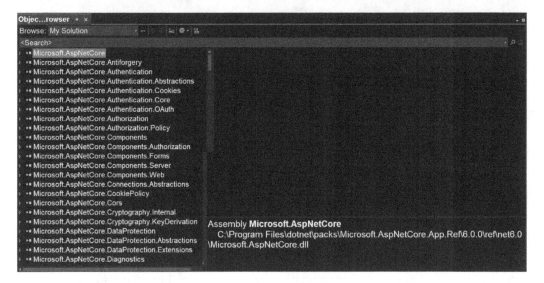

Figure 7.14 – The Object Browser window

As shown in *Figure 7.15*, at the top there is a filter, which we can deploy to choose which framework or set of libraries we want to examine:

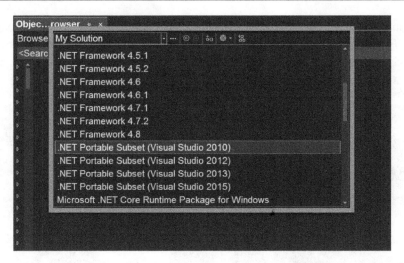

Figure 7.15 – Displaying the object browser options to navigate between the different classes

This list (in the preceding screenshot) will vary, according to the workloads picked in the Visual Studio installation.

You will also have a powerful search engine available, in which you can enter a search term, and it will return all possible matches, including any type of data that has been found, as shown in *Figure 7.16*:

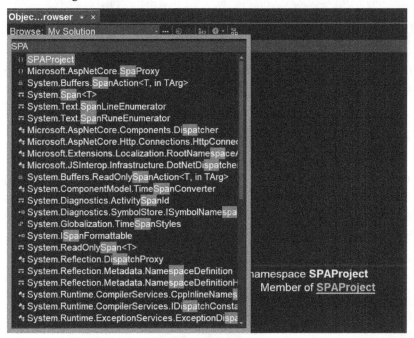

Figure 7.16 – Performing a search in the object browser

Finally, if you select any item from the list, you will see a second list on the right panel, as shown in *Figure 7.17*, which will contain all the members of the selected type, such as their methods, properties, structures, enumerations, among other available types, and at the bottom, a description of the selected member:

Figure 7.17 – Showing the members of a class in the object browser

As you can see, this window is quite helpful, not only to see members of your own project but also of the whole framework in general.

Now, let's look at a new feature called IntelliCode, which will allow us to write code more efficiently.

Using Visual IntelliCode

IntelliCode is the tool integrated into Visual Studio 2022, which allows you to write code faster, thanks to artificial intelligence. It is a tool that has been trained with thousands of popular open source projects hosted on GitHub, and although it was already beginning to show a little of its potential in Visual Studio 2019, it is in this version where all the features have been implemented.

IntelliCode can suggest patterns and styles while you write code, giving you accurate suggestions according to the context in which you find yourself, so you can complete lines of code. IntelliCode is also able to show you the methods and properties you are most likely to use and supports completion in multiple programming languages, such as the following:

- C#
- C++
- XAML
- JavaScript
- TypeScript
- Visual Basic

Let's examine how this amazing tool works in the following subsections.

Whole line completions

IntelliCode can be extremely useful in helping you complete entire lines of code. Best of all, code predictions are displayed according to different entries in your code, such as the following:

- Variable names
- Function names
- IntelliSense options used
- Libraries used in the project

There are two ways to receive entire line completion hints in Visual Studio 2022:

- The first one is given automatically while you are writing code. In *Figure 7.18*, we can see this in action, as we start writing a new property in the WeatherForecast.cs file of type string:

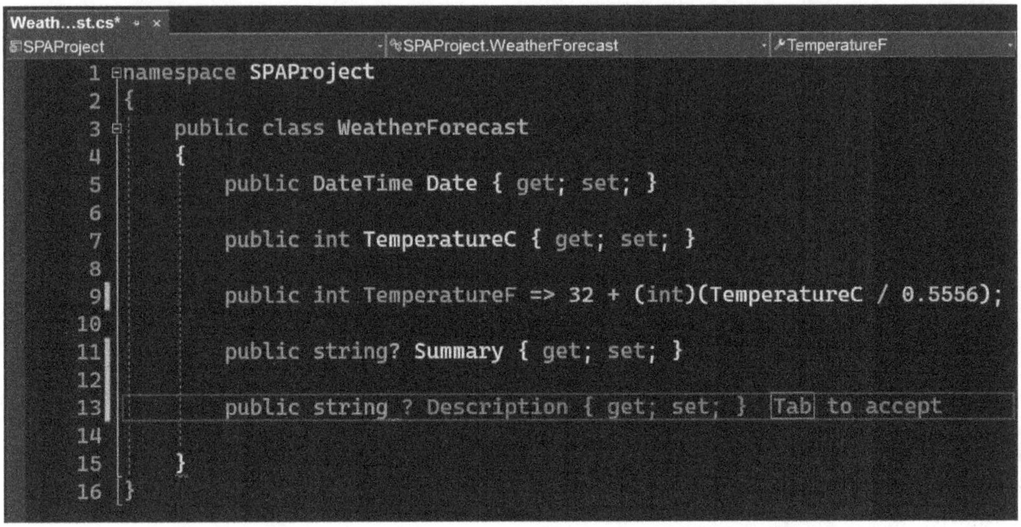

```
Weath...st.cs*  •  ×
SPAProject                              SPAProject.WeatherForecast              TemperatureF
 1  namespace SPAProject
 2  {
 3      public class WeatherForecast
 4      {
 5          public DateTime Date { get; set; }
 6
 7          public int TemperatureC { get; set; }
 8
 9          public int TemperatureF => 32 + (int)(TemperatureC / 0.5556);
10
11          public string? Summary { get; set; }
12
13          public string ? Description { get; set; }  [Tab] to accept
14
15      }
16  }
```

Figure 7.18 – IntelliCode suggesting a full line completion

According to everything learned by the IntelliCode model, it suggests a new property called Description, which we can accept by pressing the *tab* key or reject by continuing to write code.

- The second method of line completion through the use of IntelliCode is by selecting an item from the IntelliSense suggestion list. For example, if we create a constructor for the WeatherForecast class and type the letter S, a list of IntelliSense suggestions will appear. We can scroll through each of them, and in most cases, IntelliCode will show us auto-completion suggestions, as shown in *Figure 7.19*:

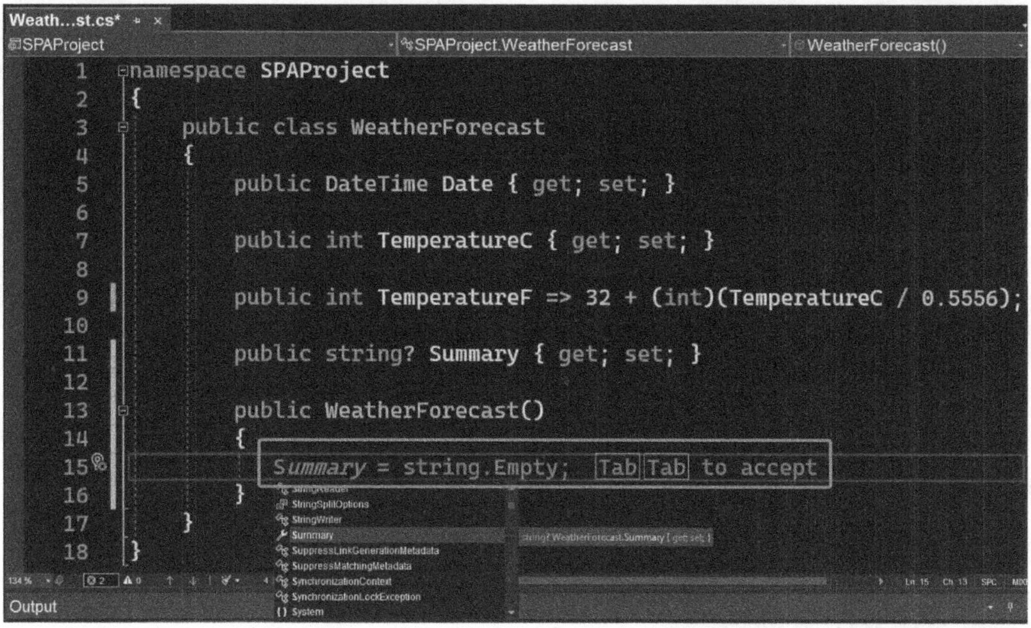

Figure 7.19 – IntelliCode recommending the completion of an IntelliSense element

We can accept the line by pressing the *tab* key twice or continue writing code to ignore the suggestion.

Now that we have seen the two methods of full line completion, let's see how IntelliCode can help us write code based on its suggestions.

IntelliCode suggestions

IntelliCode suggestions are an assisted way to carry out similar code edits in our projects. Basically, IntelliCode keeps track of code we are writing, and if it detects code repetition that could be applied to our code, it will let us know through suggestions.

A surprising thing about IntelliCode is that it is based on the semantic structure of code, so it can also help us to detect changes that we might have missed, such as changes in formulas.

For example, suppose we have some methods that allow us to calculate some static values, such as the following example:

```
public float Calculate1()
{
    var minValue = 25;
    return (float)((minValue + 126) * (Math.PI / minValue));
```

```
}

public float Calculate2()
{
    var minValue = 88;
    return (float)((minValue + 126) * (Math.PI / minValue));
}
```

We see that the calculation follows the same structure, and the only thing that changes is the value of the `minValue` variable, so we decide to create a new method called `Calculate`, which will perform the same operation by receiving a parameter:

```
public float Calculate(int value)
{
    return (float)((value + 126) * (Math.PI / value));
}
```

Subsequently, we decided to replace code of the `Calculate1` method to invoke the newly created `Calculate` method:

```
public float Calculate1()
{
    var minValue = 25;
    return Calculate(minValue);
}
```

If we go to the `Calculate2` method and start typing the name of the `Calculate` method, an IntelliCode hint will appear, as shown in *Figure 7.20*:

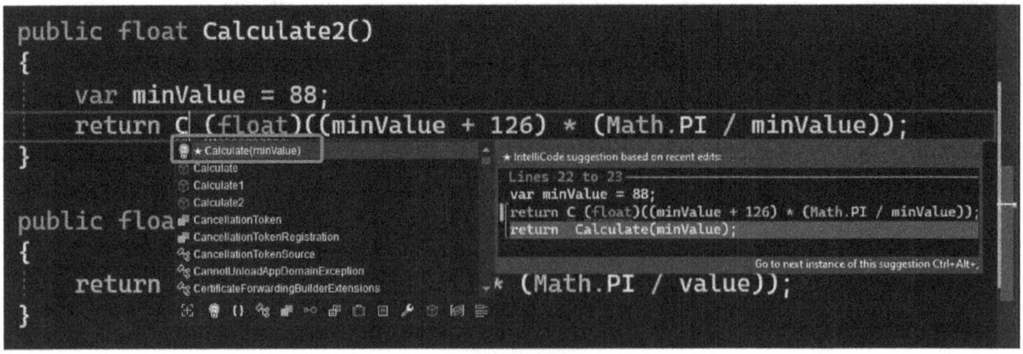

Figure 7.20 – IntelliCode suggesting an implementation of the code repetition

The suggestion made by IntelliCode tells us that we can apply the same invocation to the new method, which we can apply by pressing the *tab* key, or we can ignore the suggestion and continue writing code.

> **Important Note**
>
> It is very important to note that IntelliCode suggestions are only available during the development session. This means that if you restart Visual Studio, the previous hints will not appear again.

As we saw, IntelliCode provides a way to write code much faster through suggestions, which can save you several minutes a day.

Summary

Visual Studio contains a set of tools and windows that can help us a lot while we are developing our projects.

We have seen how CodeLens can help us find references and relationships, both through code and visually. Likewise, we have studied the different code windows that help to examine class members in projects. Finally, we have seen how IntelliCode is a new addition to the IDE, which through artificial intelligence helps us to write code quickly through various suggestions.

In *Chapter 8*, *Web Tools and Hot Reload*, we will see several tools focused on web development for the development of web applications, and a new functionality included in Visual Studio 2022 that helps us to reload a web project after making a change.

8
Web Tools and Hot Reload

Visual Studio has many tools to work with .NET applications and all the Microsoft ecosystem, but it also has many tools for other programming languages and technologies. This includes web development technologies such as JavaScript, CSS, and HTML.

In Visual Studio 2022, there are also some new improvements that help us to code faster on the frontend side. This means design tools for web developers that work with CSS, JavaScript, and HTML. With these tools, you don't need to use other editors or IDEs to complete your activities while working with these technologies.

In this chapter, you will learn about web tools in Visual Studio, and how to take advantage of them and simplify when we are developing some common statements. These tools will help you to generate code automatically, install and specify a version of web libraries, inspect the code in JavaScript, and refresh the application automatically to see the changes in real time.

We will learn the following topics related to web tools:

- Using scaffolding
- Installing JavaScript and CSS libraries
- Debugging in JavaScript
- Hot Reload

We will start using scaffolding, which is the main tool for creating project files, using templates that are included by default in Visual Studio 2022.

Technical requirements

To complete the demos of this book chapter, you must have previously installed Visual Studio 2022 with the web development workload, as shown in *Chapter 1, Getting Started with Visual Studio 2022*. It's important to have the SPA base project that was created in *Chapter 4, Creating Projects and Templates*.

You can check the changes completed in this chapter at the following link: `https://github.com/PacktPublishing/Hands-On-Visual-Studio-2022/tree/main/Chapter08`.

Using scaffolding

Scaffolding is one of the most beneficial features for developers in Visual Studio. By using scaffolding, we can save time generating code automatically by just clicking on some options.

It's indispensable to clarify that scaffolding is a popular concept in software development, and this is not unique to Visual Studio. Normally, scaffolding is associated with the code generation of **model view controller** (**MVC**) components. MVC is a popular pattern for creating web applications. Using MVC, you must distribute the responsibilities for creating web applications into three different components:

- **Model**: Responsible for saving the data

- **View**: The interface that interacts with the user

- **Controller**: In charge of handling all the actions performed by the user in the view

To use scaffolding in Visual Studio, you can select the folder in the solution explorer and right-click it to open the options to select **New Scaffolded Item…** (see *Figure 8.1*):

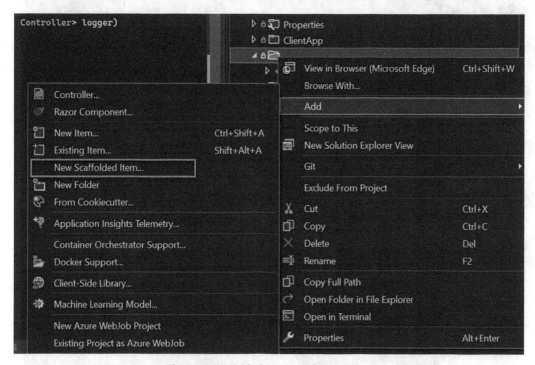

Figure 8.1 – The New Scaffolded Item… option in the project menu

Using this option, we have the possibility to create new elements in the project related to the MVC structure.

After clicking on **New Scaffolded Item…**, Visual Studio will provide a list of elements that we can create using the scaffolding tool:

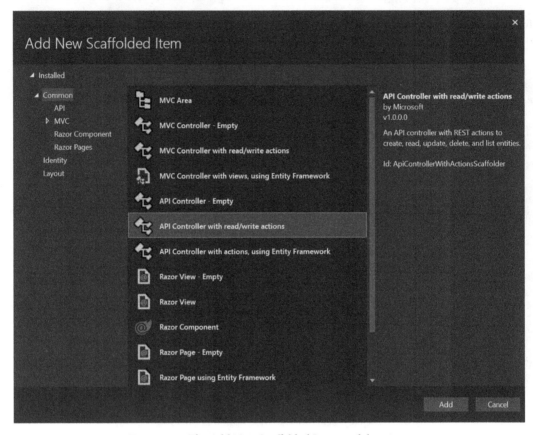

Figure 8.2 – The Add New Scaffolded Item modal options

In this case, we will select **API Controller with read/write actions,** which is going to generate an API controller with the actions for the GET, POST, PUT and DELETE verbs. Choose the name GeneratedController.cs and click on **Add**:

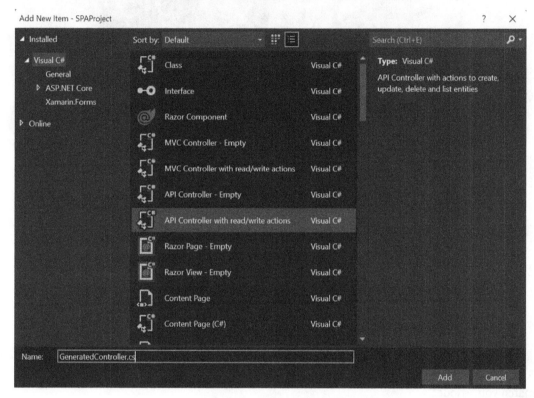

Figure 8.3 – Creating an API Controller with read/write actions

Visual Studio is going to generate a new controller in the `Controllers` folder with endpoints by default:

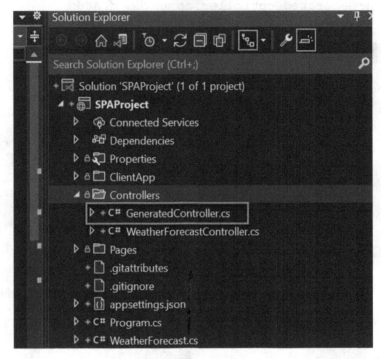

Figure 8.4 – GeneratedController is created in the Controllers folder

After creating the base template, you can replace the data type and method names to match your model.

Let's create another example of a view page using scaffolding. Using the **Add New Scaffolded Item** option again, we will select **Razor View** (see *Figure 8.5*):

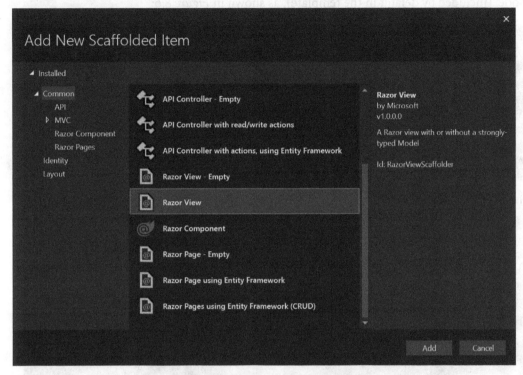

Figure 8.5 – The Razor View option in Add New Scaffolded Item

There are some templates that we can pick, including **Empty**. In this case, you should select **Create**, uncheck the **Use a layout page** option, and select the **WeatherForecast** model (included by default in the template), as shown in *Figure 8.6*:

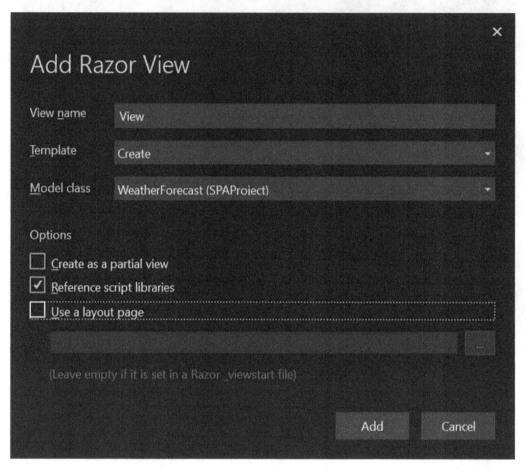

Figure 8.6 – Configuration to create a new view page using scaffolding

We can complete the creation process by clicking the **Add** button, and finally, we will see this new view page in the Pages folder. Visual Studio will analyze the model and then create a new form for each property in the model, considering the property type (see *Figure 8.7*):

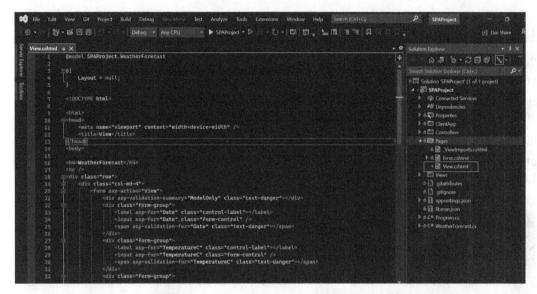

Figure 8.7 – A view created using scaffolding and the WeatherForecast model

In the View.cshtml file, we have a template for creating a new item from the **WeatherForecast** model. Scaffolding adds a label, an input, and span for control errors, as shown in this example:

```
<div class="form-group">
    <label asp-for="Date" class="control-label"></label>
    <input asp-for="Date" class="form-control" />
    <span asp-validation-for="Date" class="text-danger">
        </span>
</div>
```

> **Important Note**
>
> Scaffolding is related to ASP.NET code; we can create controllers and views but not JavaScript components.

Now you know how to use scaffolding in your projects and save time using some base templates provided by Visual Studio. Let's see another tool in Visual Studio that helps us to include JavaScript and CSS libraries in our projects.

Installing JavaScript and CSS libraries

To start a project, we can use a template from Visual Studio to easily create a **proof of concept** (**POC**), demo, or base project, but there is a probability of the project growing in functionalities and services. In this scenario, we will have to include libraries to potentialize and optimize our project and extend the functions incorporated in the base template.

To include a new JavaScript library in our SPAProject, you can open **Solution Explorer** and right-click on the ClientApp folder. In the menu, you will find the **Client-Side Library...** option (see *Figure 8.8*):

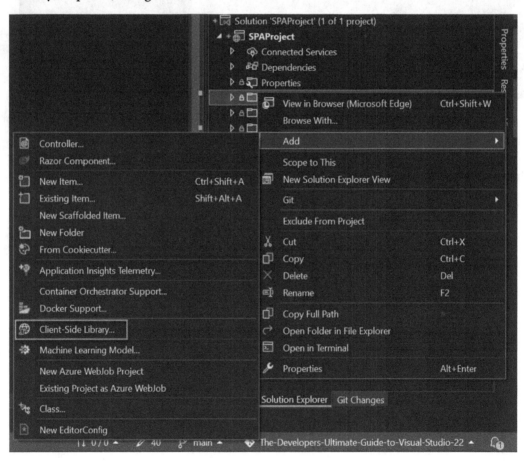

Figure 8.8 – The Client-Side Library… functionality in Visual Studio

After clicking on this option, you will get a modal that allows you to include web libraries from different resources. By default, **cdnjs** is selected, but you can also choose the other sources supported by Visual Studio:

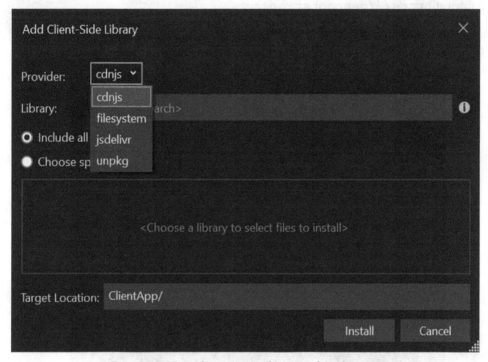

Figure 8.9 – Providers supported by Visual Studio 2022

There are three public and trusted libraries for CSS and JavaScript that we can include in our projects. Visual Studio has support for different library sources; let's review them:

- **cdnjs**: Fast and reliable content delivery for an open source library supported by Cloudflare
- **filesystem**: Custom packages in our local system
- **jsdelivr**: Free content delivery network integrated with GitHub and npm
- **unpkg**: Global open source content delivery maintained by Michael Jackson

> **Important Note**
>
> You can create the `libman.json` file manually, include the libraries in the json file, and then install them using Visual Studio. For more information, you can check `https://docs.microsoft.com/aspnet/core/client-side/libman/libman-vs`.

To analyze how Visual Studio adds a library to your project, you should select **cdnjs** and search by **bootstrap.** Bootstrap is a powerful library to create web interfaces easily using CSS classes. For more information, you can check the official documentation and quick-start guides at `https://getbootstrap.com/`.

Once you start typing, Visual Studio will suggest a list of libraries to which the written keyword is related (see *Figure 8.10*):

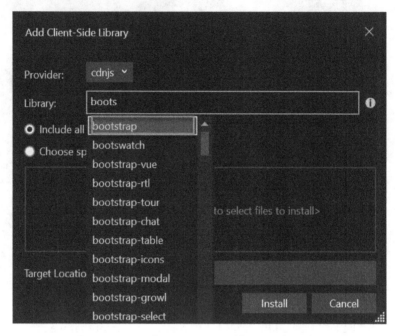

Figure 8.10 – Libraries suggested by Visual Studio

After selecting **bootstrap** (the first option recommended in the list), the most recent version of this library will be selected by default – `bootstrap@5.1.3` (see *Figure 8.11*). You can choose all the components associated with the library, but normally you only use the minify version. You can select only the files that you need, using the **Choose specific files** option. In *Figure 8.11*, you can see the `bootstrap.min.js` file, which is the only file required to use this library:

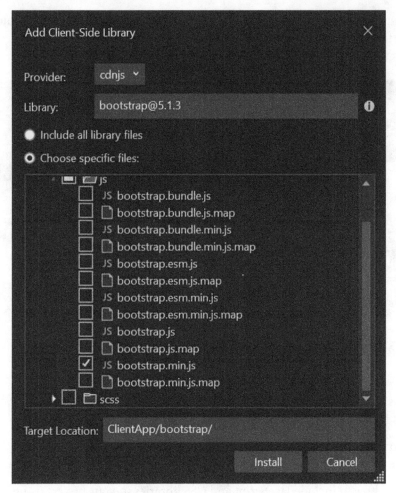

Figure 8.11 – The Bootstrap library selected and the bootstrap.min.js file picked

Now, you can click on **Install** to include Bootstrap in our project.

After installing this library, you will see a new folder that contains all the files related to Bootstrap in the `ClientApp` folder. Also, you will see a new file called `libman.json` that contains the libraries installed in the project using Visual Studio:

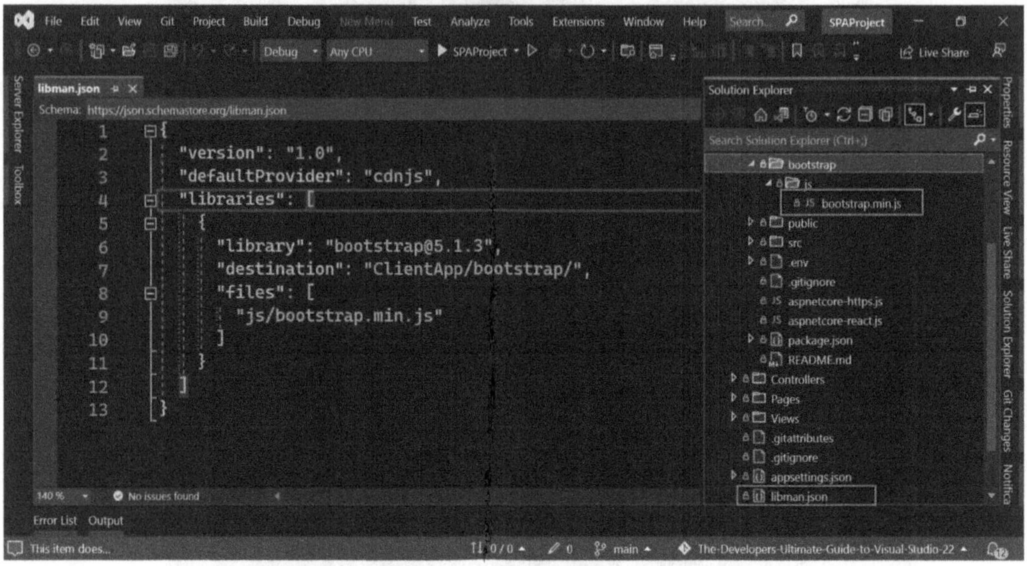

Figure 8.12 – The Bootstrap library added in SPAProject

This file helps Visual Studio to get the libraries from the servers when the project doesn't have the files for these libraries in the repository.

Visual Studio will execute all the processes automatically and create the `libman.json` file, where we can see the version of each library and the destination folder in the project.

You now know how to include JavaScript and CSS libraries in your project using the different options supported by Visual Studio. Let's learn how to debug JavaScript code in Visual Studio to find and resolve issues quickly in the development process.

Debugging in JavaScript

We must debug a project when there is strange behavior, an issue, or a blocker in our application. Visual Studio supports debugging for many programming languages, including JavaScript. This is a great feature, given that we can debug the frontend side (for example, with JavaScript) and the backend side (for example, with C#) using the same IDE.

To debug JavaScript and TypeScript code using Visual Studio, we need to check the **Script Debugging (Enabled)** option. This option is on the execution menu of the project:

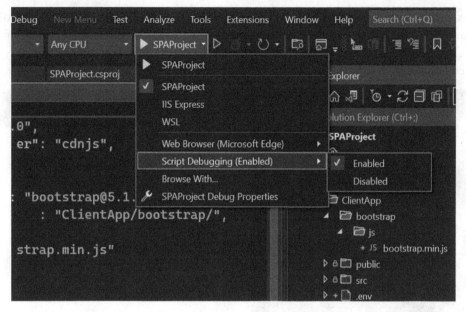

Figure 8.13 – The Script Debugging (Enabled) option in the project execution menu

Then, we can run the project in debug mode, but before that, we need to add a break to inspect the code. Navigate to `ClientApp | src | components | Counter.js` and create a new breakpoint in line number 13 (see *Figure 8.14*):

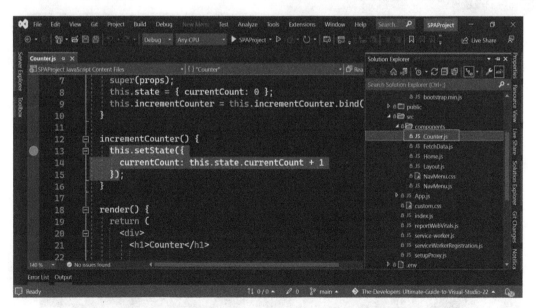

Figure 8.14 – A breakpoint in the incrementCounter method inside Counter.js

Now, execute the project using the option in the banner or press *F5* and then navigate to the counter module. Once you click on **Increment**, Visual Studio will stop the execution in the `incrementCounter` method in JavaScript. In *Figure 8.15*, you can see this expected behavior:

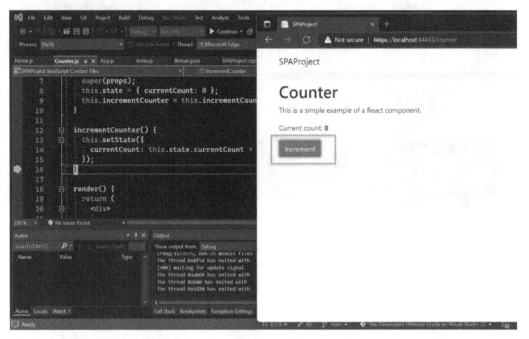

Figure 8.15 – The debugging process in the incrementCounter method with the breakpoint

At this point, we can inspect the variables in this file – for example, in *Figure 8.16*, we can see the values of `Counter.name` and `currentCount`:

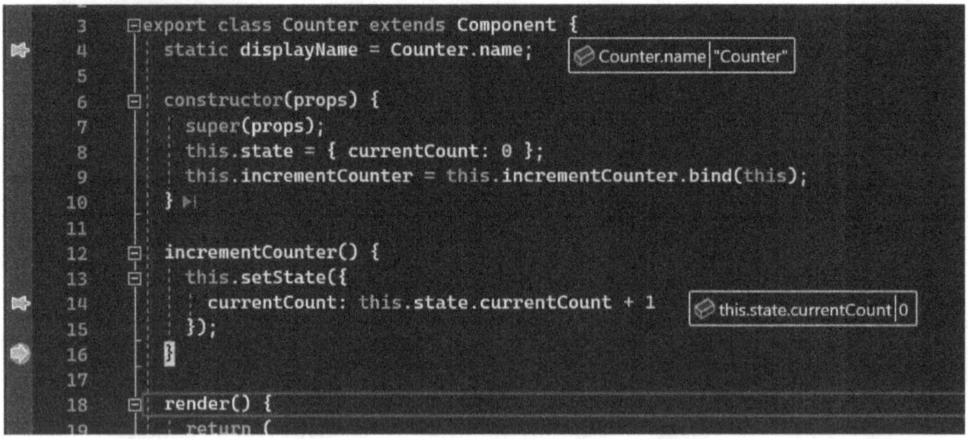

Figure 8.16 – The inspection of Counter.name and currentCount during debugging

`Counter.name` equals `"Counter"` and `currentCount` equals 0. After executing the `incrementCounter` async method, the variable in the `currentCount` state will have a value of 1.

You can use the options to execute the lines of code step by step related to the workflow for inspecting the variables, as shown in *Figure 8.17*:

Figure 8.17 – Step Over and tools for the debug process

You can now debug JavaScript code in Visual Studio and use the same tools and actions that we reviewed in *Chapter 5, Debugging and Compiling Your Projects*.

In the next section, we will review a new functionality in Visual Studio 2022 to refresh the UI after performing changes in the code.

Hot Reload

For many years, C# developers were waiting for a feature that would allow them to see real-time changes in web applications. The big challenge with this was the naturalness of C# as a programming language because C# is a compiled language. A compiled language needs to be converted to a low-level language for use by an interpreter, and this process consumes time and resources in a machine. In *Figure 8.18*, you can see a new flame-shaped icon. After clicking on this icon, you will refresh the changes in the browser, or you can select the **Hot Reload on File Save** option to reload a web application automatically after saving changes:

Figure 8.18 – The Hot Reload option in Visual Studio enabled during the execution

The Hot Reload feature has some settings that we can modify according to our needs. You can access the features using the **Settings** option when the Hot Reload button is enabled, or you can navigate to **Tools | Options | Debugging | .NET / C++ Hot Reload**:

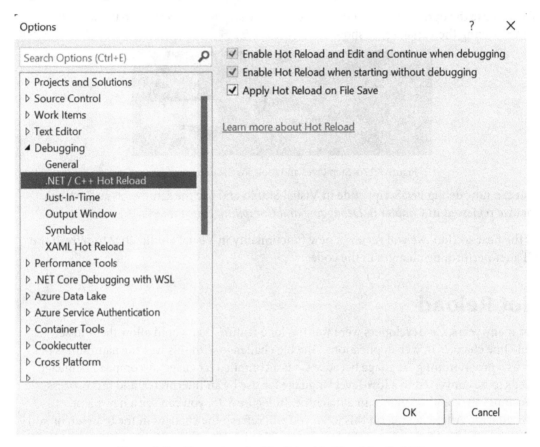

Figure 8.19 – The settings for Hot Reload in Visual Studio 2022

Let's review these options:

- **Enable Hot Reload and Edit and Continue when debugging**: This option enables Hot Reload in debug mode.

- **Enable Hot Reload when starting without debugging**: This option enables hot reload without debugging.

- **Apply Hot Reload on File Save**: After performing a change in any file and saving it, the application will reload.

> **Note**
>
> If you do not enable the **Apply Hot Reload on File Save** option, you need to use the hot reload button to refresh the web application and see the changes.

To test this functionality in Visual Studio, you can run a project by pressing *F5* or using the option in the **Debug | Start Debugging** menu. After this, you can make any change in the UI – for example, you can navigate to the `NavMenu.js` file and change the name of `NavLink` from `Counter` to `Counter Module`:

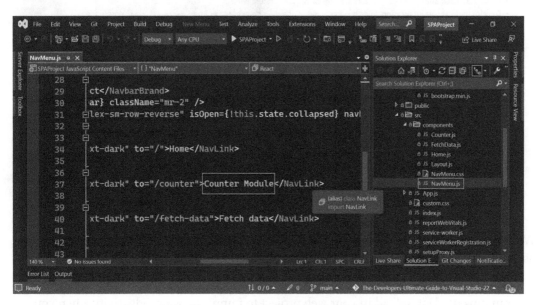

Figure 8.20 – The Counter Module NavLink in NavMenu.js

After saving the file using *Ctrl + S* or the **File - > Save All** menu, you will see the change in real time in the web browser. Check the changes shown in *Figure 8.21*:

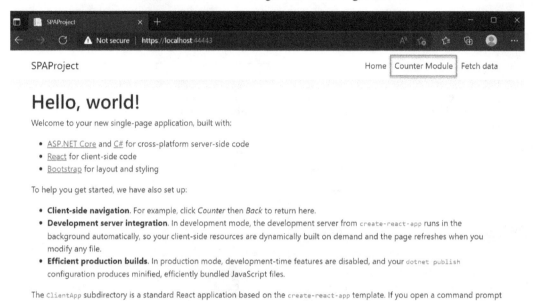

Figure 8.21 – Checking the changes in the UI related to Counter Module

---

Important Note

The wait time to see the changes in the browser depends on the project size and machine resources. Normally, it doesn't take more than 2 or 3 seconds.

We don't need to include a library in the project or install an extension in Visual Studio to use this amazing functionally. Hot reload is supported for all the web projects, including ASP.NET MVC, SPA with React and Angular (such as our SPAProject), and Blazor.

You can now use Hot Reload in your web project to improve your productivity when you are coding and need to test the changes in the UI quickly.

This brings us to the end of this chapter. Let's now review what we learned.

Summary

Now, you are ready to take advantage of the web tools in Visual Studio to code faster and improve the quality of your code. With scaffolding, we have the possibility to create components for an MVC model easily. Visual Studio generates the code using a template with simple sample code.

You also learned how to include JavaScript and CSS libraries using the tools included in Visual Studio. Using these tools, you know how to select the right version of the library and upgrade dependencies in the future.

If there is an issue or strange behavior in the code, you can now use JavaScript debugging to analyze the code deeper and execute the code step by step, inspecting the values of the variables and workflows.

Finally, you learned how to use hot reload in Visual Studio 2022 to refresh the application when you are debugging and see the changes performed in the code in real time.

In *Chapter 9, Styling and Cleanup Tools*, we will continue learning about tools included in Visual Studio that improve our experience working with styles and CSS. We will also use some tools to clean up code by choosing a specific file or a whole project.

9
Styling and Cleanup Tools

Whether you are a frontend developer or a backend developer, having tools that help you maintain clean code is essential to move projects forward in an efficient manner.

Similarly, having tools that allow you to edit CSS files quicker makes a development team move faster on a project. Fortunately, Visual Studio has several tools that you can use while working with CSS files that will allow you to write and complete your styles quickly and in a user-friendly way.

If you are a backend programmer and you use C# or Visual Basic, you should also know that there are code analysis tools, both to maintain good quality and to follow nomenclature that you can define.

These are the main topics we will cover in this chapter:

- Working with CSS styling tools
- Cleaning code with code analysis tools

Technical requirements

To perform the tests explored throughout this chapter, you must have installed the workload shown in *Chapter 1, Getting Started with Visual Studio 2022*.

In addition, to follow along with the *Working with images* section, you must install an additional component called **Image and 3D model editors**, as shown in the following figure:

Figure 9.1 – The Image and 3D model editors individual component selection

You can find the pieces of code that have been added to the project in the following repository:

```
https://github.com/PacktPublishing/Hands-On-Visual-
Studio-2022/tree/main/Chapter09
```

Woking with CSS styling tools

Having tools for editing CSS files is an advantage for frontend web developers, as it allows them to edit these files in a fast and easy way. That is why Visual Studio contains several tools that can be of great help in creating and editing these files.

Let's start by examining CSS3 snippets.

CSS3 snippets

Even today, there are still cross-browser compatibility issues for the display of styles. Surely it must have happened to you that when implementing a CSS property, it looks different on each of the browsers.

It is for this reason that Visual Studio has implemented a CSS3 snippet completion system that allows cross-browser compatibility without having to write code for each browser.

To see this in a practical way, we can open the `SPAProject | ClientApp | src | components | NavMenu.css` file and locate the `.box-shadow` style. Within this style, we can start typing the term `border-radius`, which will display the list of IntelliSense recommendations, as shown in *Figure 9.2*:

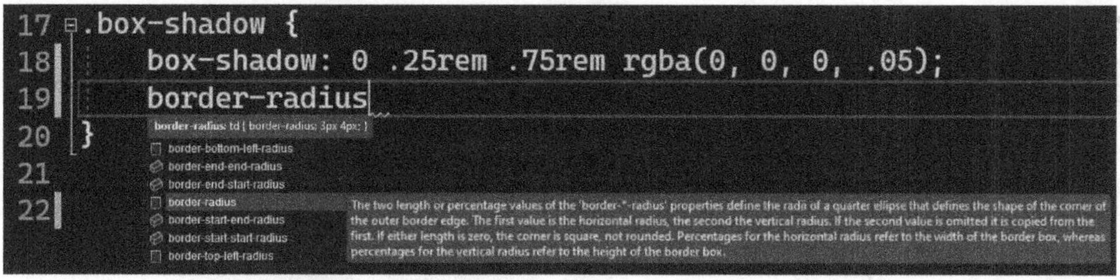

Figure 9.2 – Recommendations for the term border-radius

You can see that there are two types of icons in this list, some with a blue geometric shape and others with a square white figure. Among these icons, we are interested in those of white color, since they are the CSS3 snippets. We can scroll through the list with the keys on the keyboard, and once we have selected the snippet we are interested in, just press the *tab* key twice, which will result in the implementation of the cross-browser-compatible CSS3 snippet, as shown here:

```css
.box-shadow {
    box-shadow: 0 .25rem .75rem rgba(0, 0, 0, .05);
    -moz-border-radius: inherit;
    -webkit-border-radius: inherit;
    border-radius: inherit;
}
```

Among the most common multi-browser CSS3 styles, we can find the following:

- Alignment styles
- Animation styles
- Background styles
- Border styles
- Box styles
- Column styles

- Flex styles

- Grid styles

- Mask styles

- Text styles

- Transition styles

As you can see, these snippets can help you in the creation of styles by attacking cross-browser compatibility in an effective way. Let's see now how Visual Studio can help us understand CSS styles more quickly through indentation.

Hierarchical CSS indentation

Style indentation is a visual aid that can increase productivity considerably by showing the content of a style through spaces at the beginning of a line, as well as the sub-styles belonging to a parent style.

Visual Studio allows you to create a quick indentation in the style files. Suppose, for example, you want to create a style called `.main` and a sub-style that affects all `div` elements within the `.main` style, as shown here:

```
.main {
    padding: 0px 12px;
    margin: 12px 8px 8px 8px;
    min-height: 420px;
}
.main div {
    border: 25px;
}
```

In principle, if you have written the styles at the same indentation level, you can apply the indentation by going to the **Edit | Advanced | Format Document** menu to perform a hierarchical indentation of the whole document, as shown in the following code block:

```
.main {
    padding: 0px 12px;
    margin: 12px 8px 8px 8px;
    min-height: 420px;
}

    .main div {
```

```
        border: 25px;
    }
```

If, on the other hand, you only want to apply the indentation on a specifically selected set of styles, you can select the **Edit | Advanced | Format Selection** option.

> **Note**
>
> It is possible to customize the indentation values through the **Tools | Options | Text Editor | CSS | Tabs** option.

Now, let's look at the color picker feature in Visual Studio.

Color picker

One feature that is extremely useful when working with styles is the ability to select a color for an element. Fortunately, Visual Studio has a built-in color picker that, although looking very simple, does its job very well.

To test it, let's edit the `.main` style we created in the *Hierarchical CSS indentation* section. Type the `background-color:` attribute, which will show you a vertical display of predefined colors with an assigned name that you can select. Ignore this list and instead type the # symbol. Immediately, you will see a new horizontal list of predefined colors, as shown in *Figure 9.3*:

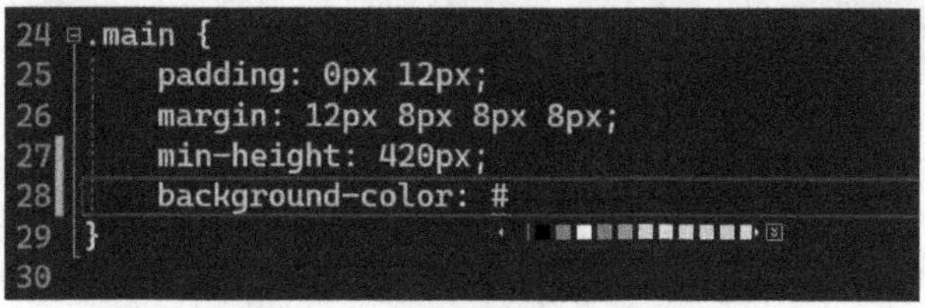

Figure 9.3 – A horizontal list of predefined colors

If you want to set a custom color, you can click on the button at the end of the color list, which will display the color picker, as shown here:

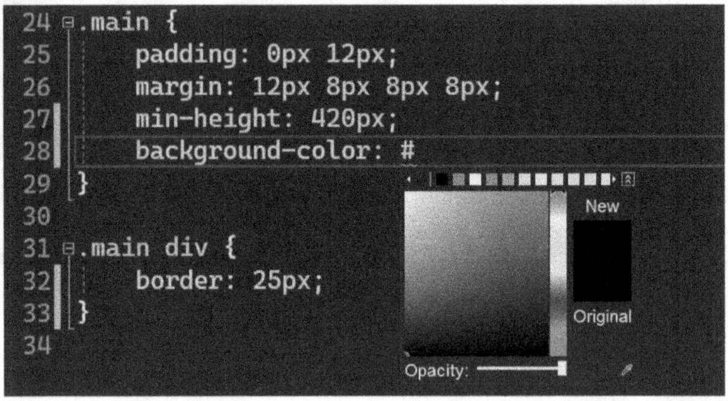

Figure 9.4 – The CSS color picker

This way, it is possible to select a color from the color selection, change the color hue, add opacity or transparency to the selected color, and even use the eyedropper tool to select the color from an external source, such as an image. For this demonstration, I have selected the color with the #1b0b8599 code, as seen in *Figure 9.5*:

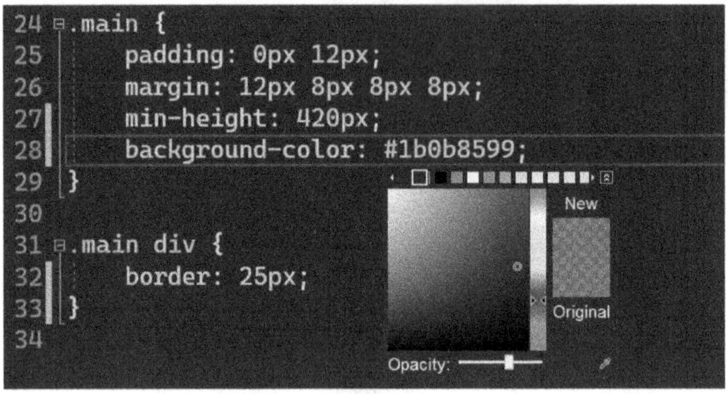

Figure 9.5 – Selecting a custom color from the color picker

A great advantage of this tool is that it will store the custom-selected colors internally, so you can reuse them over and over again throughout your CSS files. For example, if we go to the .main div style and want to assign a custom color to the color attribute, we will see that in the horizontal color list, the custom color that we have used in *Figure 9.5* is listed, as shown here:

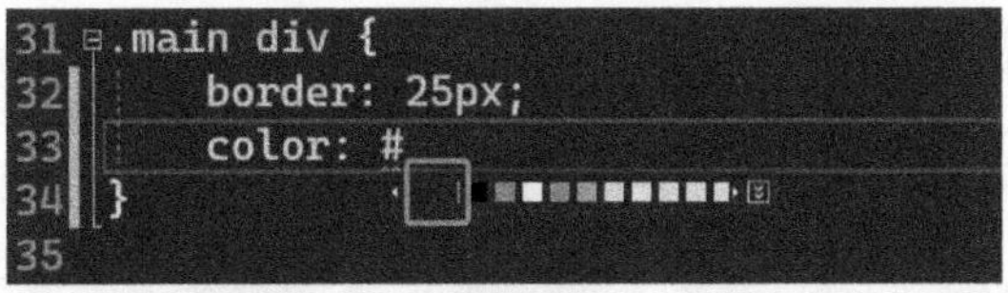

```
31 ⊟.main div {
32        border: 25px;
33        color: #
34 }
35
```

Figure 9.6 – A stored list of previously used custom colors

The color picker, without a doubt, is a tool that can help us a lot when we need to assign a custom color. Now, let's find out how IntelliSense can help us write faster in CSS files.

IntelliSense in style files

Just as IntelliSense can be an extraordinary help when creating source code, it can also be very useful when creating style files.

Let's look at some practical examples of IntelliSense usage. Let's go to the `ClientApp` | `src` | `components` | `NavMenu.css` file. Inside this file, let's proceed to create a new style called `.intellisense`, as we can see here:

```
.intellisense{

}
```

If we position ourselves inside the style and press the *Ctrl + spacebar* key combination, the list of all the attributes that we can add to the newly created style will be displayed. If we start typing the name of an attribute, it will start filtering the list with the matches of what we type, as shown in *Figure 9.7*:

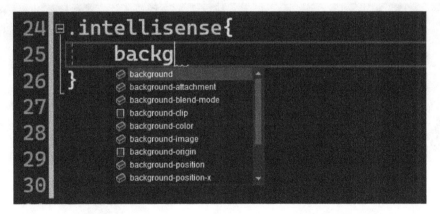

Figure 9.7 – IntelliSense showing recommendations

Also, it is possible to select an item from the list and complete the item name by pressing the *tab* key.

For this demonstration, let's assume that we need to use the background attribute, but we do not know the possible values we can assign to it. IntelliSense can help us in a visual way by showing us an example of the use of each attribute, as we can see in the following figure:

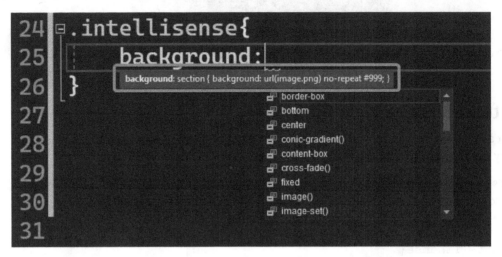

Figure 9.8 – IntelliSense showing a possible use of the background attribute

Not only that, but IntelliSense also adapts the results to the context of the selected attribute. For example, suppose we need to assign a set of fonts to the font-family attribute of a style. If we type the font-family attribute, Visual Studio will provide us with the list of values corresponding to the font-family attribute, as shown in *Figure 9.9*:

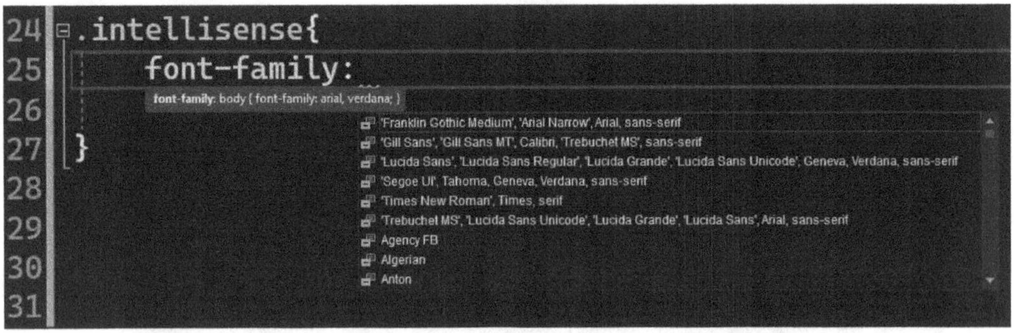

Figure 9.9 – Recommended values for the font-family attribute by IntelliSense

If, on the other hand, we want to assign a value to the font-weight attribute, we will see results according to this attribute, as shown in *Figure 9.10*:

Figure 9.10 – Recommended values for the font-weight attribute by IntelliSense

Undoubtedly, IntelliSense is an excellent aid for the creation of styles. Now, let's move on to analyze the image editor.

Working with images

A tool that is not very well known in Visual Studio is the image editor. This tool must be installed as specified in the *Technical requirements* section, and without a doubt, it can help us in the basic editing of the images of our project.

Here are some of the situations in which the image editor is useful:

- When we need to rescale an image
- When we need to change the color of a section to another color
- When we need to rotate an image
- When we need to add text to an image
- When we need to apply a filter to an image

In the repository mentioned in the *Technical requirements* section, I have added an image located at SPAProject | ClientApp | public | visualstudiologo.png to perform different tests with the image editor.

Once we open an image (in this case, `visualstudiologo.png`), we will see two toolbars – one located on the left side, called the *image editor* toolbar, and the second one on the top, called the image editor mode toolbar, as shown in the following figure:

Figure 9.11 – The Visual Studio image editor

First, let's analyze the image editor toolbar. This is a bar that appears on the left side of the editor and contains tools that allow you to perform some action on the image, such as adding geometric shapes or rotating the image.

At the top, we have the image editor mode toolbar. This toolbar contains buttons that execute advanced commands, such as irregular selection, wand selection, pan, zoom, and image properties.

Let's look at a practical example. Suppose we need to execute the following tasks on the image:

- Convert the image to grayscale.
- Flip the image horizontally.
- Write the text `Visual Studio Logo` on the image.

To execute these tasks, we must perform the following steps in order:

1. In the image editor mode toolbar, select the **Advanced | Filters | Black and White** option, as shown in the following figure:

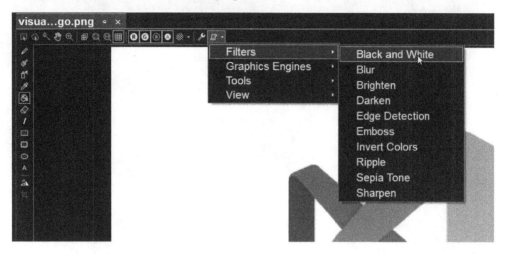

Figure 9.12 – Converting the image to grayscale

2. In the image editor toolbar, double-click the rotate image button, as shown here:

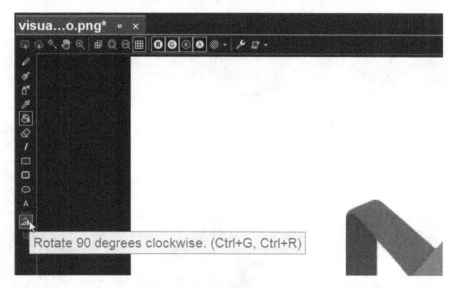

Figure 9.13 – Rotating the image

3. Select the text tool, as shown in *Figure 9.14*:

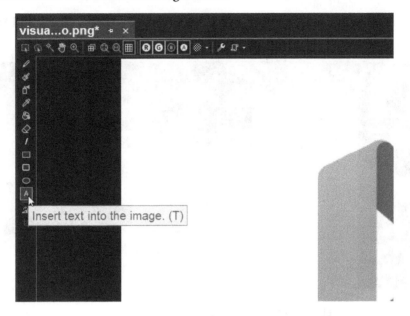

Figure 9.14 – Selecting the text tool

4. Add the text `Visual Studio Logo` in the **Properties** window, as shown in *Figure 9.15*:

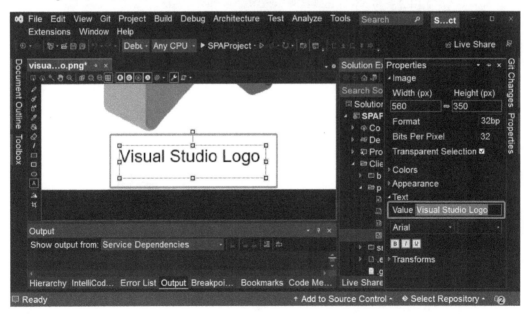

Figure 9.15 – Changing the Text value in the Properties window

With these edits applied, we will have the result shown in *Figure 9.16*:

Visual Studio Logo

Figure 9.16 – The result after applying the changes

The last thing to do is to save the image so that the changes are permanently applied to it. With this example, we have been able to see how the image editor can be very useful if we need to make edits to our images.

In the next part, let's understand how code analysis can help .NET developers to have clean and quality code.

Cleaning code with code analysis tools

Visual Studio 2022 includes a series of C# or Visual Basic code analyzers that allow us to maintain good code quality and consistent style in the source code. To use this feature by default, projects must be configured on a framework version of .NET 5 or higher. To differentiate compilation errors, analysis violations will appear with the **CA** prefix in the case of a code quality analysis violation and **IDE** in the case of a style analysis violation.

The code analysis tools correspond to code quality analysis and code style analysis, so we will see what they are, how they can help us, and learn how to set up and run with code cleanup profiles. Let's first look at how to take advantage of code quality analysis.

Code quality analysis

Code quality consists of having a source code that is secure, with the best possible performance and good design, among other characteristics. Fortunately, Visual Studio can help us to maintain high-quality code through rules enabled by default.

To visualize one of these violations in a practical way, let's go to the `Program.cs` file and add the following line at the end of the file:

```
int value1 = 1;
int value2 = 1;
Console.WriteLine(Object.ReferenceEquals(value1, value2));
```

Now, to compile the project, right-click on the project name and click on the **Build** option, as shown in the following figure:

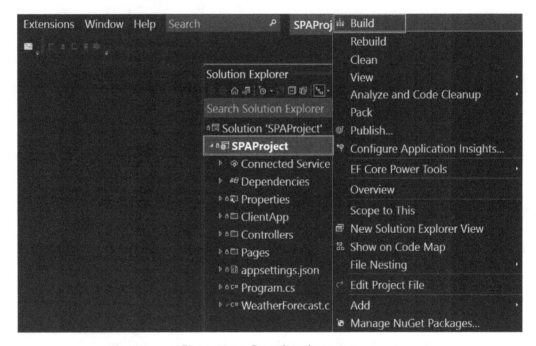

Figure 9.17 – Compiling the project

When performing the compilation, we do not see errors immediately; however, if you go to the **Error List** tab and see the list of **Warnings**, you will see some marked with the **CA** prefix. In our specific example, we can see the **CA2013** warning, as shown in *Figure 9.18*, which tells us not to pass a value of type `int` to the `ReferenceEquals` method because it will always return a *false* value due to the *boxing* operation (conversion from a *value* type to a *reference* type) of the value:

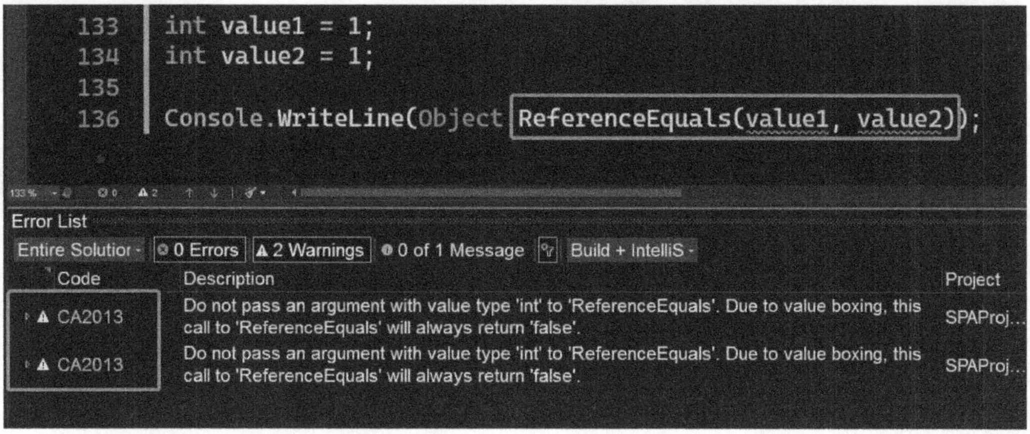

```
133    int value1 = 1;
134    int value2 = 1;
135
136    Console.WriteLine(Object ReferenceEquals(value1, value2));
```

Error List

Entire Solution ▾ | ⊘ 0 Errors | ⚠ 2 Warnings | ● 0 of 1 Message | 🔍 Build + IntelliS ▾

Code	Description	Project
⚠ CA2013	Do not pass an argument with value type 'int' to 'ReferenceEquals'. Due to value boxing, this call to 'ReferenceEquals' will always return 'false'.	SPAProj…
⚠ CA2013	Do not pass an argument with value type 'int' to 'ReferenceEquals'. Due to value boxing, this call to 'ReferenceEquals' will always return 'false'.	SPAProj…

Figure 9.18 – A code quality warning

> **Important Note**
> Occasionally, Visual Studio may suggest code fixes to fix warnings in the code, through a light bulb icon appearing on the error. Also, you can see the complete list of code quality rules at the following URL: `https://github.com/dotnet/roslyn-analyzers/blob/main/src/NetAnalyzers/Core/AnalyzerReleases.Shipped.md`.

Let's now look at the rules applied to the code style.

Working with code styles

Code styles are configurations that can be quite useful for C# and Visual Basic developers to keep a project with correct nomenclature, especially if the project is used by several members of a team.

Code styles can be created for a specific project or Visual Studio instance installed on a machine.

The way to use code styles is by opening the **Tools | Options | Text Editor | C# or Visual Basic | Code Style | General** section:

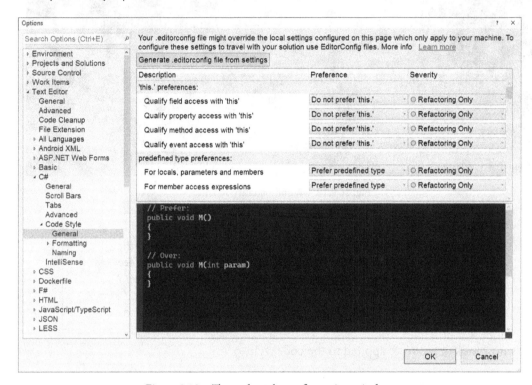

Figure 9.19 – The code style configuration window

Once we are in this window, we will be able to see the code style configuration for the current machine. We can change any of these options to fit the code nomenclature we need.

If we need to specify a configuration file to be applied as part of the solution, even if it is opened on another machine, we can modify the configuration values. Once we have the settings we want to follow throughout the solution, we should click on the **Generate .editorconfig file from settings** button, as shown here:

Your .editorconfig file might override the local settings configured on this page which only apply to your machine. To configure these settings to travel with your solution use EditorConfig files. More info   Learn more

Generate .editorconfig file from settings

Description	Preference	Severity
'Me.' preferences:		
Qualify field access with 'Me'	Do not prefer 'Me.'	○ Refactoring Only
Qualify property access with 'Me'	Do not prefer 'Me.'	○ Refactoring Only
Qualify method access with 'Me'	Do not prefer 'Me.'	○ Refactoring Only
Qualify event access with 'Me'	Do not prefer 'Me.'	○ Refactoring Only
Predefined type preferences:		
For locals, parameters and members	Prefer predefined type	○ Refactoring Only
For member access expressions	Prefer predefined type	○ Refactoring Only

Figure 9.20 – The button to generate a configuration file

This will open a dialog, asking for the name and path where the configuration file will be saved. In this example, it has been saved as `config.editorconfig`, as shown in the following figure:

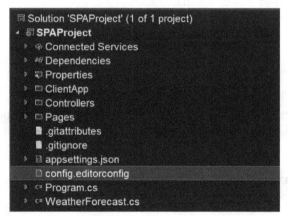

Figure 9.21 – The configuration file created

If we proceed to open the file we have created, the text editor will open. Here, we will be able to see the applied configuration in text format, being able to change the preselected parameters quickly. Once again, these changes will accompany the solution so that all source files that are part of the project and have the same code nomenclature.

Now, let's talk about code cleanup profiles that will control which aspects should be applied when code is cleaned.

Configuring a code cleanup profile

The code cleanup profiles are a configuration in which you can indicate what type of code cleanup you want to apply to your project. There are several ways to access the profile configuration window, but the general way is to go to the **Tools | Options | Text Editor | Code Cleanup | Configure Code Cleanup** menu. This will display the following figure:

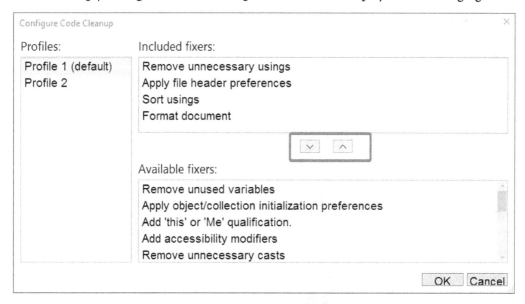

Figure 9.22 – The Configuration Code Cleanup window

As you can see, two cleaning profiles can be configured with different options, with **Profile 1** the one that will be executed by default. Likewise, inside the window we have two sections:

- The **Included fixers** list contains the specific active actions that we want to apply throughout the project.

- The **Available fixers** list contains the actions that are currently disabled, but at any time, we could add them to the active actions.

Fixers can be enabled or disabled easily, with the arrow buttons marked in *Figure 9.22.*

Executing code cleanup

Once we have created the code cleanup file and established the fixers that will be applied for the cleanup, let's see how we can apply this cleanup.

To do this, we are going to the bottom of the editor, where we will press the button with the broom icon, as shown in *Figure 9.23*:

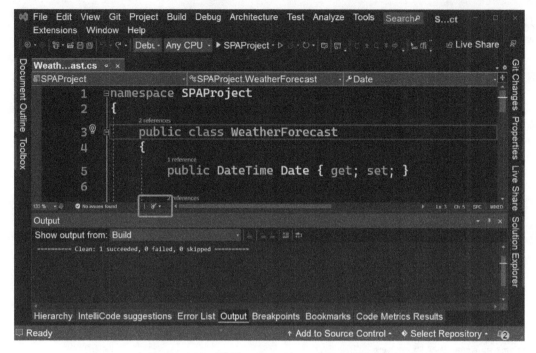

Figure 9.23 – The button to execute code cleanup

As mentioned in the *Configuring a code cleanup profile* subsection, this will apply only the rules configured in the active cleaning profile. For example, suppose we have the following configuration in the default configuration profile:

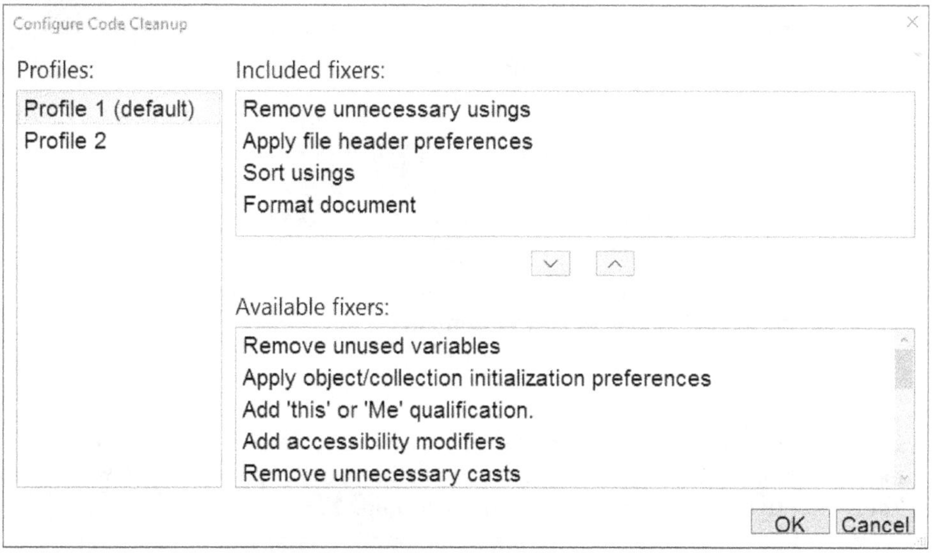

Configure Code Cleanup

Profiles:

Profile 1 (default)
Profile 2

Included fixers:

Remove unnecessary usings
Apply file header preferences
Sort usings
Format document

Available fixers:

Remove unused variables
Apply object/collection initialization preferences
Add 'this' or 'Me' qualification.
Add accessibility modifiers
Remove unnecessary casts

OK   Cancel

Figure 9.24 – A set of selected test fixers

Then, we decide to modify the `WeatherForecast.cs` file with the following code without indentation, and with an unused `using System.Data.Common` namespace:

```
using System.Data.Common;
namespace SPAProject
{
public class WeatherForecast
{
 public DateTime Date { get; set; }
 public int TemperatureC { get; set; }
 public int TemperatureF => 32 + (int)(TemperatureC /
     0.5556);
 public string? Summary { get; set; }
}
}
```

Now, when we apply code cleanup, it will result in clean code, as shown here:

```
namespace SPAProject
{
    public class WeatherForecast
    {
        public DateTime Date { get; set; }
        public int TemperatureC { get; set; }
        public int TemperatureF => 32 + (int)
            (TemperatureC / 0.5556);
        public string? Summary { get; set; }
    }
}
```

Undoubtedly, this tool can be of great help to maintain a consistent and clean style, whether we work individually or with a development team.

Note

It is possible to configure Visual Studio to perform a code cleanup every time a file is saved through the **Run Code Cleanup profile on Save** option, located in **Tools | Configuration | Text Editor | Code Cleanup**.

Summary

In this chapter, we have learned about the different tools that Visual Studio has for frontend and backend developers.

We learned how CSS3 snippets can help create cross-browser compatible styles quickly. Likewise, hierarchical CSS indentation helps to keep styles readable. Also, the color picker can help to select colors quickly, and we have also seen how IntelliSense is present when we need to edit CSS files, and finally, how the image editor provides useful tools if we need to make basic edits to our images.

In the case of code analysis, we have learned how code quality analysis can help us to have safe and reliable code, while code style analysis helps us to maintain a nomenclature whether we are working individually or with a team of developers.

In *Chapter 10*, *Publishing Projects*, you will learn the most common ways to publish web projects on different platforms.

10
Publishing Projects

After finishing a **proof of concept** (**POC**) or a **minimum viable product** (**MVP**), (which means a demo or pilot project with the main functionalities implemented, as discussed in *Chapter 8, Web Tools and Hot Reload*), we need to deploy our changes to see how the project works in a real scenario and share the published project with our customers. Visual Studio has a set of tools to deploy our projects. We can choose an option to deploy the project in our local environment, but we can also use services in the cloud.

In this chapter, you will learn how to deploy your project with just a few clicks and Visual Studio 2022. These tools will help you to save time and reduce complexity when you need to deploy.

We will discuss and review the following topics, which are the options to publish projects with Visual Studio 2022:

- The options to publish a project
- Publishing in a folder
- Publishing in **internet information services** (**IIS**)
- Publishing in Microsoft Azure

Let's dive in and learn all about publishing projects.

Technical requirements

To publish a project following the steps provided in this chapter, you must have previously installed Visual Studio 2022 with the web development workload, as shown in *Chapter 1, Getting Started with Visual Studio 2022*. It's important to have the SPA base project that you created in *Chapter 4, Creating Projects and Templates*.

For the *Publishing in Microsoft Azure* section, you will need to have an Azure account with credits to complete the deployment.

You can check out the changes made to the project at the following link: `https://github.com/PacktPublishing/Hands-On-Visual-Studio-2022/tree/main/Chapter10`.

The options to publish a project

Visual Studio has two ways to access the **Publish** option and then display the possibilities incorporated to deploy a project.

In the menu, we can access this option by navigating to **Build** | **Publish [Project name]**. By default, the main project in the current solution is selected, as shown here:

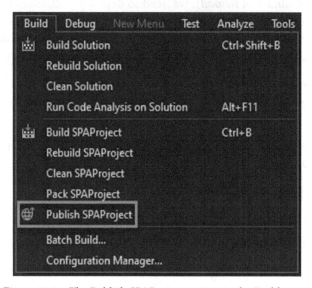

Figure 10.1 – The Publish SPAProject option in the Build menu

Another possibility is to right-click on the project that we want to publish in the **Solution Explorer** tab (see *Figure 10.2*):

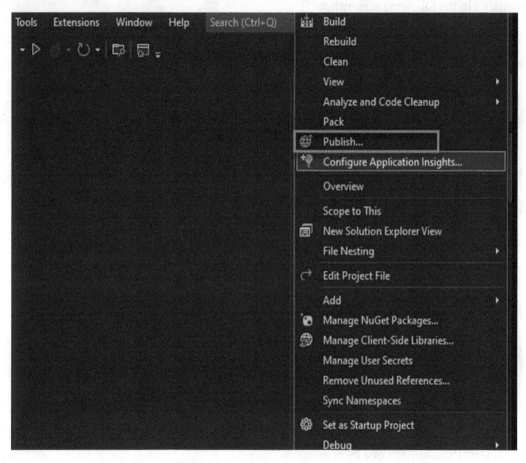

Figure 10.2 – The Publish… option on the menu after right-clicking

Whatever the chosen option, Visual Studio will display a modal window to give you the deployment types supported for the main project in the solution or the project selected, as illustrated in the following screenshot:

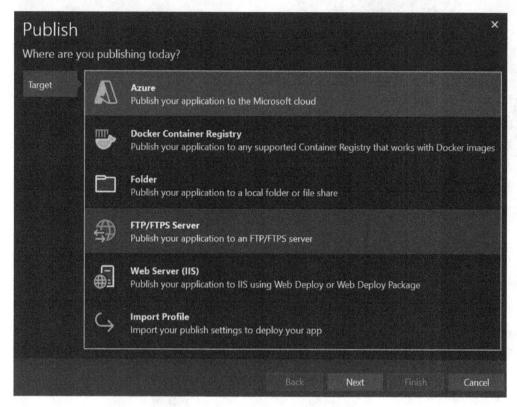

Figure 10.3 – The options to publish a project in Visual Studio

Let's review these options one by one:

- **Azure**: The project will be deployed in Azure using a service, depending on the technology used.

- **Docker Container Registry**: This option helps us publish a Docker container, including our website.

- **Folder**: This option publishes the project in a folder chosen in the filesystem.

- **FTP/FTPS Server**: With this option, the files of the publication will be transferred to an FTP server.

- **Web Server (IIS)**: Deploys the project in IIS, the default server in Windows.

- **Import Profile**: Allows us to import another configuration to publish a project already created.

We will review some of these options in the following sections. We will convert the project with the **Azure**, **Folder**, and **Web Server (IIS)** options because **Docker Container Registry** is a little more complex to set up, and for FTP/FTPS server, we need to have an FTP server running, which is not very common in modern solutions.

At this point, you know how to access the **Publish** option that Visual Studio contains to deploy a web application. Let's now explore a few preferred options to publish a project in Visual Studio.

Publishing in a folder

One of the most common options for publishing a project is to use our filesystem and save the site in a folder, including all the resources to use it in a local server, such as IIS, Apache, or NGINX. Visual Studio has an option to publish our projects with this approach easily.

You can use any option to navigate to the publishing functionality, which we reviewed in the *The options to publish a project* section, and then select **Folder**, followed by a click on **Next**:

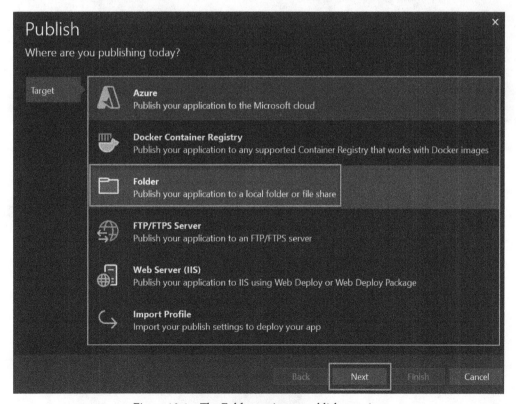

Figure 10.4 – The Folder option to publish a project

In the next window, we can specify the folder where we want to save the published project using the **Browse...** option. We can use absolute and relative paths. Then, we can finish the configuration by clicking on **Finish**:

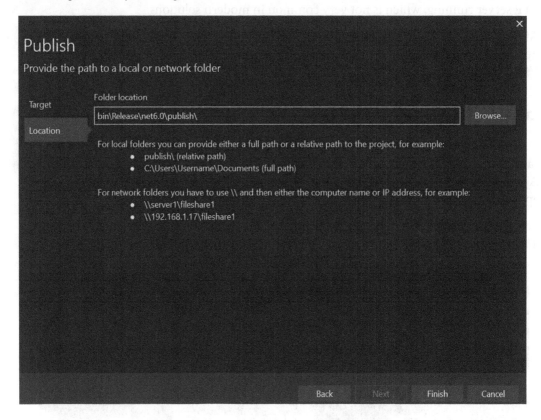

Figure 10.5 – The folder location to publish the project

Important Note

The folder selected must be empty. If you select a folder that contains files, this could create conflicts in the files. Visual Studio will try to replace the files with the same name using the files generated in the publishing process.

After completing the publishing configuration, Visual Studio will generate a file with the
`.pubxml` extension that contains the options we chose before in XML format. Now, we
can use the **Publish** button to publish the project in the selected folder:

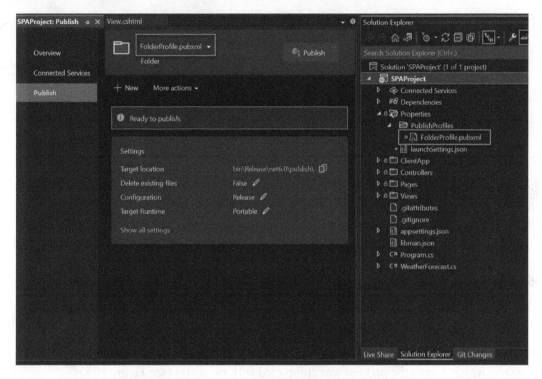

Figure 10.6 – The .pubxml file created with the configuration

After clicking on **Publish**, Visual Studio will show a **Publishing to Folder…** message and a console log, where you can check all the steps performed during this process. In *Figure 10.7*, you can see the log in the **Output** panel:

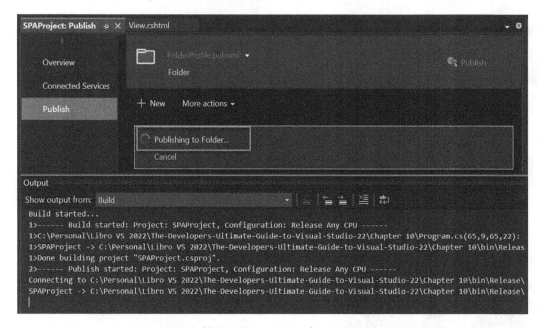

Figure 10.7 – Publishing in progress after clicking on Publish

Finally, after waiting for some seconds, we will see a **Publish succeeded** confirmation message, which means the process was completed with no issues (see *Figure 10.8*):

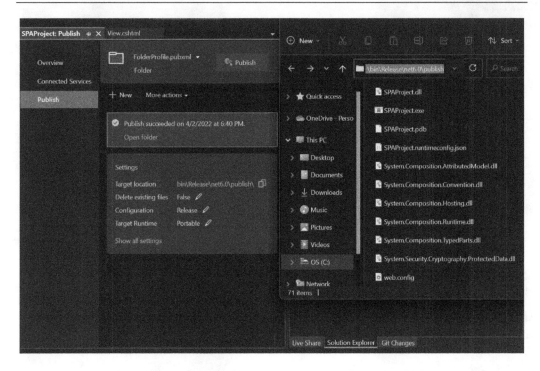

Figure 10.8 – The publish success message and the publish folder

In *Figure 10.8*, we can see that the folder on the right side contains all the files associated with the publication of our projects.

We have some additional settings that we can change using the pencil icon:

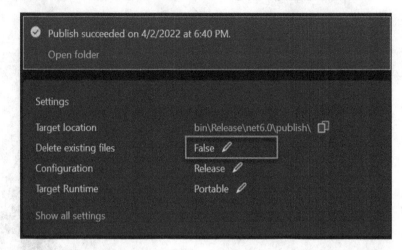

Figure 10.9 – The option to delete an existing file in the target location

After clicking the **Delete existing files** option, we will see some additional settings that we can change according to our needs (see *Figure 10.10*):

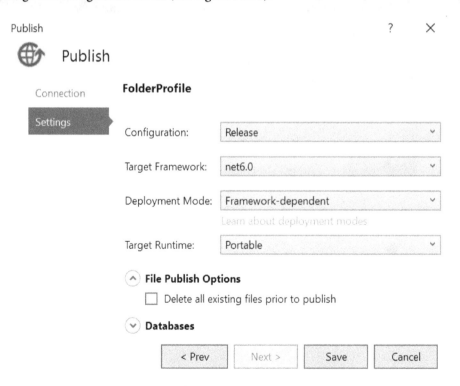

Figure 10.10 – Additional settings to publish a project

Let's review all the options in *Figure 10.10*:

- **Configuration**: We can choose between **Debug** and **Release**. The latter is optimized for production.

- **Target Framework**: Normally, it's the same version used in the project, but we can select a specific version to run our project.

- **Deployment Mode**: We can select **Framework-dependent**, which means the framework should be installed on the server, or **Self-contained**, which means the publication will include the framework and all the dependencies to run the project.

- **Target Runtime**: This is the system architecture and operating system where the application will be published.

- **Delete all existing files prior to publish**: This allows Visual Studio to delete all the existing files in the folder selected for publishing the project.

> **Important Note**
> It's best practice to check the **Delete all existing files prior to publish** option to avoid conflict with existing files during the publishing process.

You now know how to publish in a folder in Visual Studio. You only need to select the folder where you prefer to deploy your application and then use the publish option.

Publishing in IIS

IIS is the most popular server in the Windows ecosystem. It is included by default in all Windows Server versions, and there is an option to install it on Windows.

To install IIS in Windows 8, 10, 11, or later, you can follow this guide in the *Step 1 – Install IIS and ASP.NET* section of the Microsoft document:

```
https://docs.microsoft.com/windows/msix/app-installer/
web-install-iis
```

You can easily deploy a web application in IIS using Visual Studio. First, you need to open the **Publish** option that we reviewed in the *The options to publish a project* section and select **Web Server (IIS)** (see *Figure 10.11*):

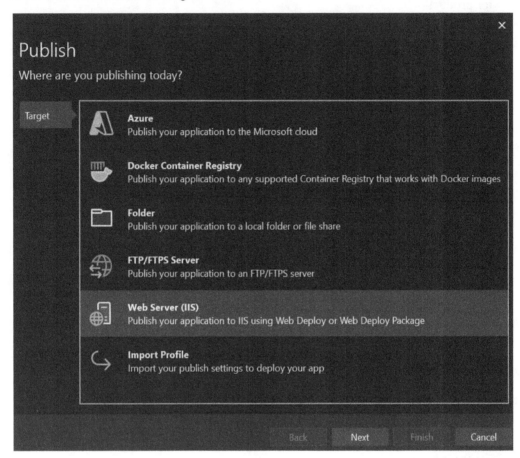

Figure 10.11 – The Web server (IIS) option to publish a project

> **Important Note**
>
> If you already have a `.pubxml` file in the publishing profiles folder, you need to delete it to create a new one with a new configuration.

After selecting **Web Server (IIS)**, you can click on **Next** and continue with the process. You will see two options to choose from (see *Figure 10.12*):

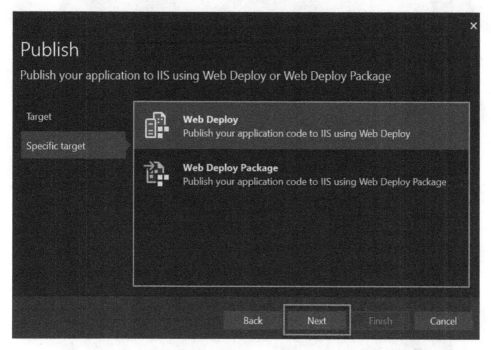

Figure 10.12 – The options to deploy in IIS – Web Deploy and Web Deploy Package

We have two options for deploying in IIS:

- **Web Deploy**: Deploy the folder, including all the files, in IIS.
- **Web Deploy Package**: Create a `.zip` file, including all the files within the publishing folder.

In this case, we will select **Web Deploy** and click on **Next**.

You need to fill in the **Server**, **Site name**, and **Destination URL** fields (see *Figure 10.13*):

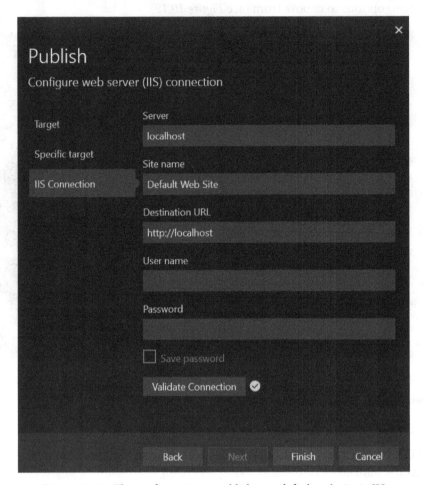

Figure 10.13 – The configuration to publish on a default web site in IIS

We will use the default website created in IIS. If you are already using this site with another application, you can set up another site name in IIS.

Using the **Validate Connection** button, you can check whether Visual Studio can create the site and complete the publication using the setup provided. You can use the **Finish** button to complete the setup.

> **Important Note**
> You need to execute Visual Studio as admin to give access to IIS; otherwise, you will receive an error.

After completing all the steps, you will get a **Ready to publish** message and some options to edit the configuration. We are ready to use the **Publish** button:

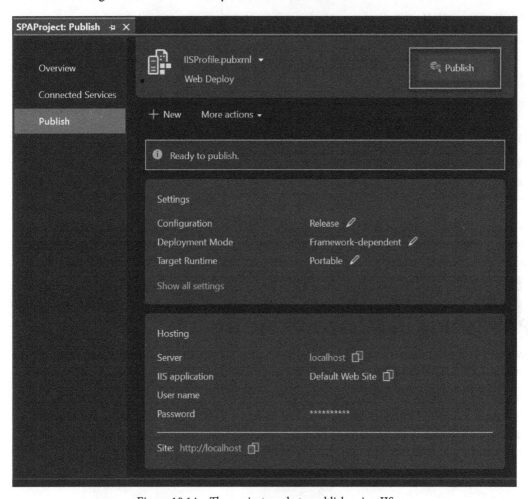

Figure 10.14 – The project ready to publish using IIS

Automatically, Visual Studio will open in a new window in the browser, with the URL of the site that we set up in the configuration (see *Figure 10.13*) in the **Destination URL** field. In this case, Visual Studio will open `http://localhost`:

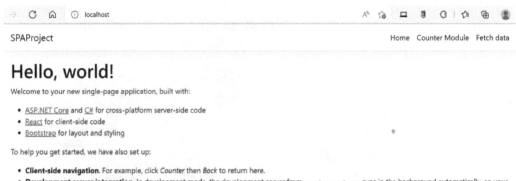

Figure 10.15 – The project running in IIS after publishing it

You just learned how to deploy a web application in IIS using Visual Studio. Visual Studio minimizes the steps to complete the setup and sets all the files in a folder inside IIS. Now, let's review another way to deploy an application using Azure.

Publishing in Microsoft Azure

Azure is a cloud provider created by Microsoft, and it's one of the most popular among start-ups and .NET developers. Since Azure and Visual Studio are supported by the same company and community, there is good integration between the technologies.

To deploy our SPAProject in Azure, we need to select a publish option using the method we reviewed in the *The options to publish a project* section, select **Azure**, and then on the next screen, select **Azure App Service (Windows)**:

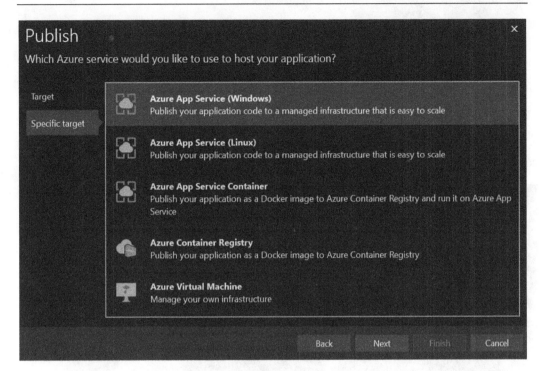

Figure 10.16 – The possibilities to publish in Azure

Remember to delete any `.pubxml` file in the project to create a new configuration.

> **Important Note**
> To complete the steps in this section and publish the SPAProject in Azure, you will need an Azure account with credits.

Visual Studio provides some options to deploy in Azure. Let's review them:

- **Azure App Service (Windows)**: Publish the project in a **platform as a service** (**PaaS**) container using Windows as the operating system.

- **Azure App Service (Linux)**: Publish the project in a PaaS container using Linux as the operating system.

- **Azure App Service Container**: Run up a container in an Azure App using an Azure App Service container.

- **Azure Container Registry**: Publish the application as a Docker image.

- **Azure Virtual Machine**: Publish a folder in a virtual machine and then manually use IIS to run the app.

After clicking on **Next**, you will get a new window to perform authentication in Azure and connect with the services and resources in Azure for your account:

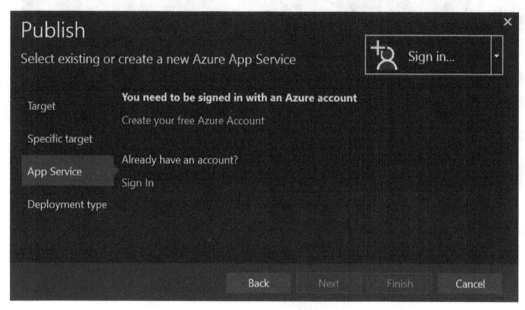

Figure 10.17 – The Azure account sign-in

In *Figure 10.17*, we see a new window, where we need to authenticate with Azure using the **Sign In…** button to continue with the setup and integrate Visual Studio with the resources in our Azure account. After performing the authentication, we can select the subscription and create a new web app instance using the green cross button (see *Figure 10.18*):

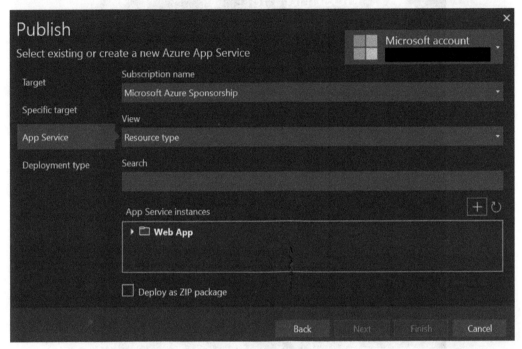

Figure 10.18 – The option to select a subscription and create a new app service

We need to select a hosting plan for our SPAProject:

Figure 10.19 – The hosting plan for a new app service

In this case, we can use a free plan in Azure with limited time to run our web app. It's perfect for performing demos and trying this publishing functionality.

> **Important Note**
>
> To analyze and compare other plans in Azure, you can navigate to the following link: `https://azure.microsoft.com/pricing/details/ app-service/windows/`.

After creating the hosting plan, we can finish the configuration by choosing a name and resource group for our project (see *Figure 10.20*):

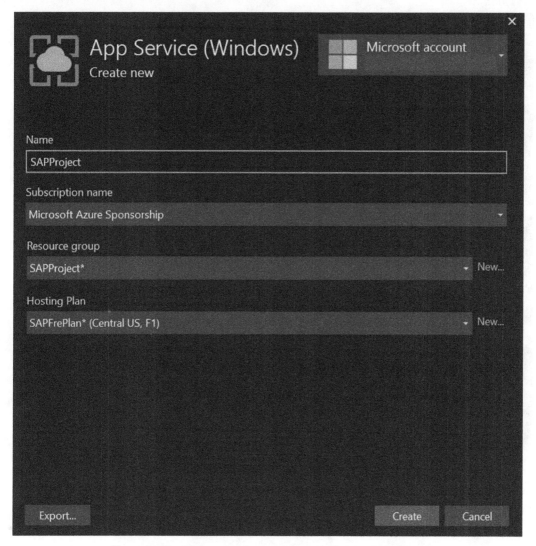

Figure 10.20 – The app service configuration for our SAPProject

In *Figure 10.21*, we can see a preview of the project created, and we can complete using the **Finish** button:

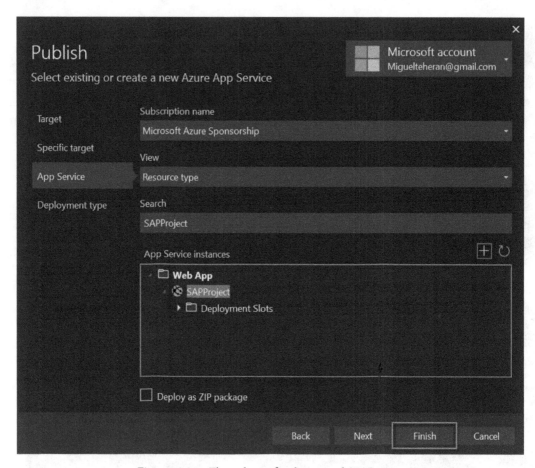

Figure 10.21 – The web app for the created SPAProject

Finally, in the last step, we have two options to complete the integration with Azure (see *Figure 10.22*):

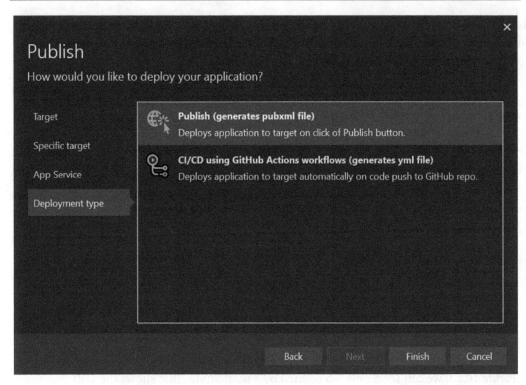

Figure 10.22 – The deployment options in Azure

Let's review these options to choose:

- **Publish (generates pubxml file)**: Create a `.pubxml` file with the configuration. We need to publish the project manually using the **Publish** button.

- **CI/CD using GitHub Actions workflows (generates yml file)**: If we have our project in GitHub, this option creates a `.yml` file that contains **continuous integration** (CI) and **continuous deployment** (CD) configurations to deploy the project after every change in code.

For this example, we will choose **Publish (generates pubxml file)** and click on **Finish** to complete the setup.

As with *Figure 10.14,* we will see the `.pubxml` file created and the **Publish** button. Click on **Publish** to publish our project in Azure. After publishing the project, Visual Studio will open the site using the URL from Azure:

Figure 10.23 – The SPAProject running using an app service in Azure

In *Figure 10.23,* we can see a new tab opened by Visual Studio that shows the URL of our project in Azure. This URL has the name of the project followed by the `azurewebsites.net` domain.

You now know how to publish a project in Azure using only the UI in Visual Studio and some clicks. To see more details about the integration between Visual Studio and Azure, check out this link: `https://docs.microsoft.com/aspnet/core/tutorials/publish-to-azure-webapp-using-vs`.

After reviewing the main options to deploy an application using Visual Studio, we conclude this chapter.

Summary

After reading this chapter, you know how to navigate in Visual Studio to access the **Publish** option, using the build menu or by accessing the project. You are qualified to use the **Folder** option and deploy your web app in the filesystem, and then use a server to host an application. You also know how to deploy a project in the IIS server using Visual Studio. You learned how to deploy a project in Azure using the app service option in Windows and a publishing configuration using the `.pubxml` file.

In *Chapter 11, Implementing Git Integration*, we will review all the tools included in Visual Studio to connect with Git repositories and GitHub in particular. You will learn how to see the status of your changes using a visual interface and publish your project in a public or private repository easily.

Part 3: GitHub Integration and Extensions

In this part, you will learn how to add more functionalities in Visual Studio extensions, how to collaborate with other developers using Live Share, and how to manage Git repositories.

This part contains the following chapters:

11
Implementing Git Integration

Having a change control platform for the development of a software project is essential for good control of the project. There are many different versioning systems, but **Git** is the most widely used system today, which is why more and more IDEs are including tools for managing repositories based on this technology natively.

This has precisely happened with Visual Studio, which integrates a series of options to allow us to work with Git-based repositories.

In this chapter, you are going to learn how to work with Git repositories based on **GitHub**, which is the most popular repository-hosting platform today.

The main topics we will see in the chapter are as follows:

- Getting started with Git settings
- Creating a Git repository
- Cloning a Git repository
- Fetching, pulling, and pushing Git repositories
- Managing branches
- Viewing changes in repositories

Technical requirements

To follow along in the chapter, you must have installed Visual Studio with the workload set from *Chapter 1*, *Getting Started with Visual Studio 2022*.

Since the projects hosted in the main repository of the previous chapters already have a GitHub configuration, we will create a new project throughout the chapter to do the exercises. Therefore, a GitHub account is required, which can be created at the following link: `https://github.com/signup`.

Getting started with Git settings

Starting to work with Git tools is very easy in Visual Studio 2022, since they are included as part of the installation itself, so you can install Visual Studio and start working on your projects as soon as possible.

To access the management of code projects hosted on GitHub, you must first sign in with a Microsoft account, as explained in *Chapter 1*, *Getting Started with Visual Studio 2022*. Once logged in, click on the account profile icon and select the **Account settings…** option, as shown here:

Figure 11.1 – Accessing the account settings

This will open an account customization window, where we can add a GitHub account via the **+Add** button, as shown here:

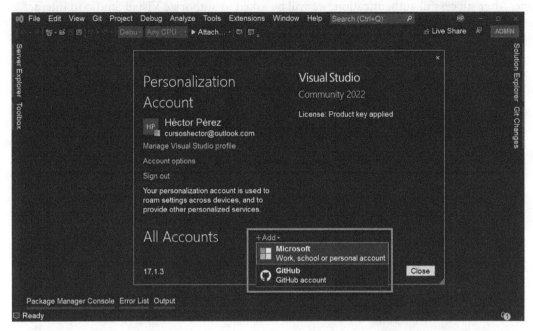

Figure 11.2 – The button to add a GitHub account

Once we press the button, we will be redirected to the GitHub authentication portal, where we can log in with an existing GitHub account or create an account if we don't have one. After successful authentication, we will be asked to authorize Visual Studio to interact with GitHub services, so it is essential to press the button that says **Authorize github**:

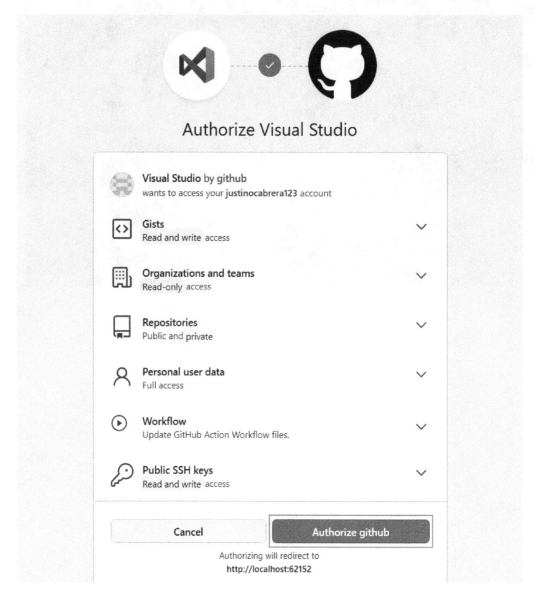

Figure 11.3 – The authorization button to connect Visual Studio and GitHub

After this step, the GitHub account will have been added as part of the accounts associated with the main Visual Studio account, so we can start working with the repositories of the GitHub account, as shown in *Figure 11.4*:

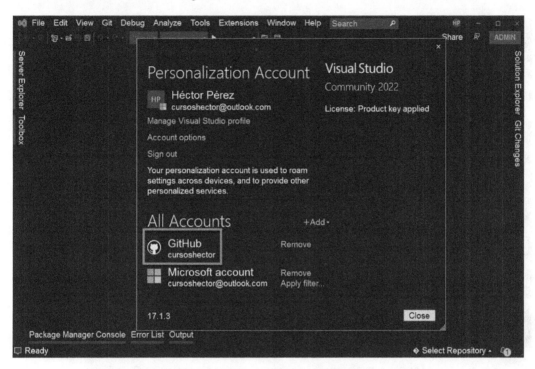

Figure 11.4 – The GitHub account added to the account listing

Now that we have associated a GitHub account, let's learn how to create Git repositories in GitHub.

Creating a Git repository

Creating a repository in GitHub from a Visual Studio project is very easy to do. In this section, you will have to test your knowledge by creating a new **ASP.NET Core Empty** project, as discussed in the *Templates for .NET Core* section of *Chapter 4, Creating Projects and Templates*, which you can name GitDemo.

To create the new repository in GitHub from the project you have created, just select the **Git | Create Git Repository** menu option and fill in the repository information, according to the data shown in *Figure 11.5*:

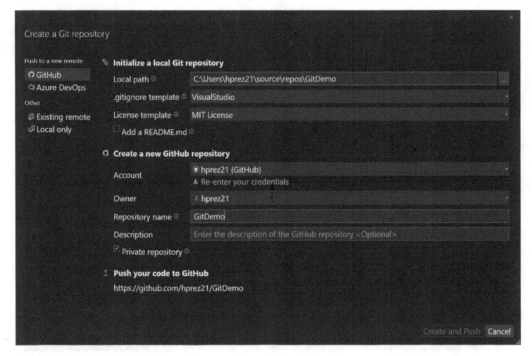

Figure 11.5 – Filling in the information for a new repository on GitHub

Let's briefly explain each option:

- **Local path**: This sets the path on the local machine where the source code is hosted (usually the path where you created the project).

- **.gitignore template**: This allows you to select a template that establishes a set of files that will not be uploaded to the repository – for example, files that are generated after a compilation and can be regenerated on a different computer are selected by the selected template.

- **License template**: This allows you to choose a license for the repository code, which indicates what users can and cannot do with the project code.

- **Add a README**: This allows you to add a `readme` file that describes the purpose of the repository.

- **Account**: This allows you to select the GitHub account to which the repository will be published. It is possible to associate an account from here if one has not been associated before.

- **Owner**: This allows you to set which GitHub account will be the owner of the repository if the account belongs to several work teams.

- **Repository name**: This allows you to set the name of the repository, although the project's own name is normally used. This will affect the URL of the final repository.

- **Description**: This allows you to enter a description of the repository created.

- **Private repository**: This allows you to set whether the repository will be public or private.

Once these values are set, just click on the **Create and Push** button to start the repository creation process. It is important that you follow this step in your GitHub account so that you can carry out the activities in the following sections.

There are several signs to identify whether a project belongs to a Git repository, as shown in *Figure 11.6*, such as lock icons on the left side of source files, which indicate that they are the original repository files:

Figure 11.6 – Signs to identify a versioned project with Git

Another sign to identify whether a project belongs to a Git repository is that (as shown in *Figure 11.6*) at the bottom, there is a branch icon, indicating which branch we are working on. In our case, the branch is **master**.

Now that we know how to create a new repository, let's see how to clone a repository to our local machine.

Cloning a Git repository

It may be the case that you want to clone an existing repository and not start from scratch because you need to work in a team, or simply while browsing the GitHub site, you came across a repository that caught your attention.

The easiest way to clone a repository is from the initial window of Visual Studio, which can be reached by either starting a new instance of Visual Studio, closing an open project, or from the **File | Start Window** menu. In this window, the first option is the **Clone a repository** button:

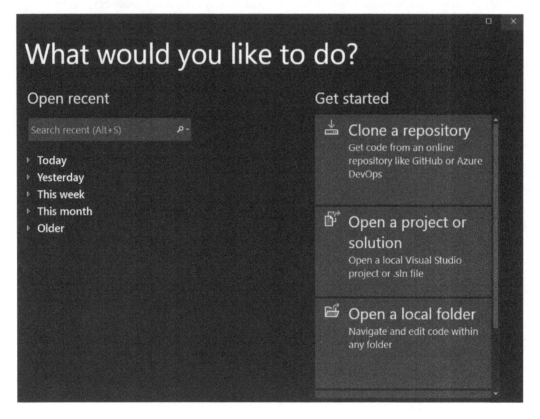

Figure 11.7 – The Clone a repository option from the startup window

Once we press this button, a new window (shown in *Figure 11.8*) will open, asking us to indicate the URL of the remote repository and the local path where the source code files will be stored:

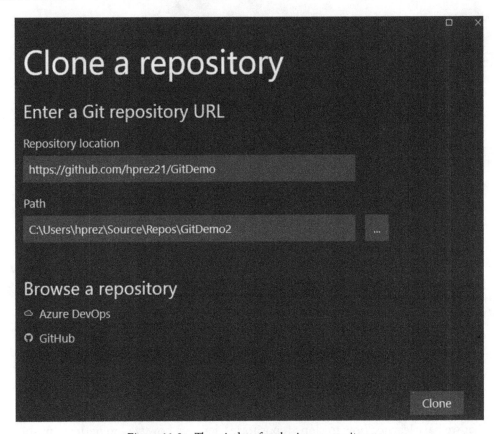

Figure 11.8 – The window for cloning a repository

For the purposes of this demonstration, we will use the URL of the repository created in the *Creating a Git repository* section, which should look like this: `https://github.com/{your-github-username-here}/GitDemo`.

Because we previously created a project with the same name in the *Creating a Git repository* section, we have to indicate a different path for the project. For simplicity, we will change the name of the folder to `GitDemo2`, as shown in *Figure 11.8*:

After filling in the information in the window, we will proceed to click on the **Clone** button, which will start the process of cloning the repository locally. Once the process is finished, the **Solution Explorer** window will show you a folder view:

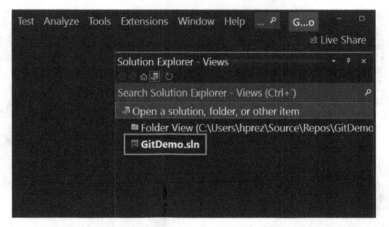

Figure 11.9 – The folder view on the Solution Explorer window

To switch from the folder view to the solution view, and to be able to work with the source files, you must open the solution by clicking on the GitDemo.sln file. This will open the solution and all its files, allowing us to work on the source files, as we have done so far.

> **Important Note**
>
> It is possible to change the default folder view to the solution view by selecting the option located at **Git | Settings | Git Global Settings | Automatically load the solution when opening a Git repository**.

Now that we have learned how to clone repositories, let's see how we can perform pushing and pulling actions on repositories.

Fetching, pulling, and pushing Git repositories

The most important commands when working with Git repositories have to do with fetching, pulling, and pushing operations. There are two main ways to execute these operations:

- The first way is by accessing through the **Git** menu, as shown here:

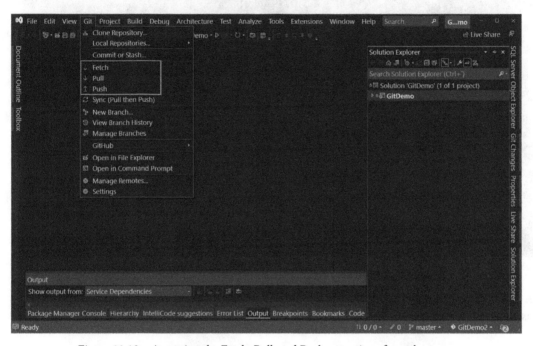

Figure 11.10 – Accessing the Fetch, Pull, and Push operations from the menu

- The second is to enable the **Git Changes** window, which you can open through the **View | Git Changes** menu, as shown in *Figure 11.11*:

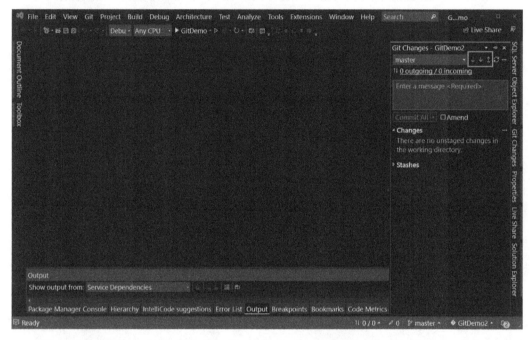

Figure 11.11 – Accessing the Fetch, Pull, and Push operations from the Git Changes window

At this point, you may be wondering what each of these operations is for. So, let's now explain them briefly.

Fetching repositories

The fetch operation allows you to check whether there are remote commits that should be incorporated into the local repository.

To run this example, go to the GitHub portal, log in if you are not already logged in, and open the repository we created in the *Creating a Git repository* section called GitDemo. Once you are in the repository, go to the Program.cs file and click on the pencil icon, which will allow you to edit the file, as shown here:

Figure 11.12 – The button to edit a repository file

We will make a very simple change by adding a pair of exclamation marks at the end of the string in line four:

```
app.MapGet("/", () => "Hello World!!!");
```

Once this change has been made, go to the bottom of the page and click on the green button to commit the changes:

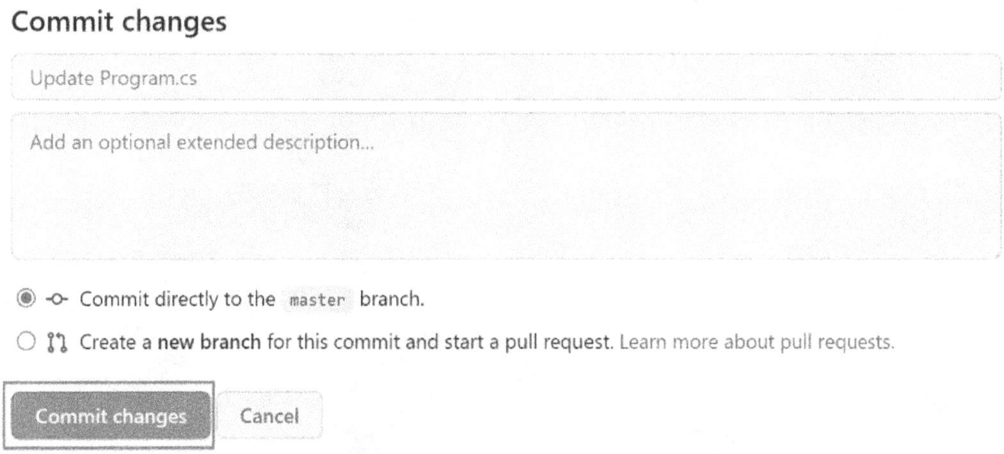

Commit changes

Update Program.cs

Add an optional extended description...

◉ ⚬ Commit directly to the `master` branch.

○ ⫙ Create a **new branch** for this commit and start a pull request. Learn more about pull requests.

[Commit changes] Cancel

Figure 11.13 – The button to commit changes on the repository

If we now go to Visual Studio and click on the **Fetch** button, as shown in *Figure 11.14*, we can see a **0 outgoing / 1 incoming** message, indicating that there is a change in the repository that has not been applied to the local project:

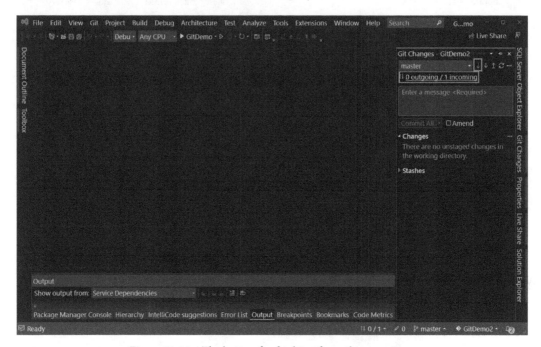

Figure 11.14 – The button for fetching from the repository

Additionally, if we click on this message, a new window will open, which will show us the version history of the project, showing us which changes we have not been applied locally:

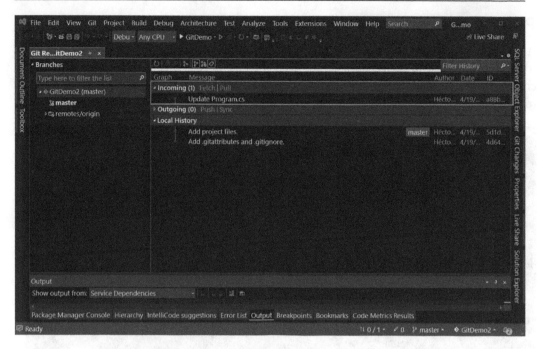

Figure 11.15 – The window showing the pending changes to be applied to the local repository

It should be noted that the fetch operation does not make any changes locally or in the remote repository, which the pull and push operations do. We will discuss these operations next.

Pulling repositories

A pull operation refers to the act of downloading the latest changes from the repository to our local project. In the *Fetching repositories* section, we saw that we have a pending change to apply. So, we will proceed to click on the **Pull** button, as shown in *Figure 11.16*, to apply it:

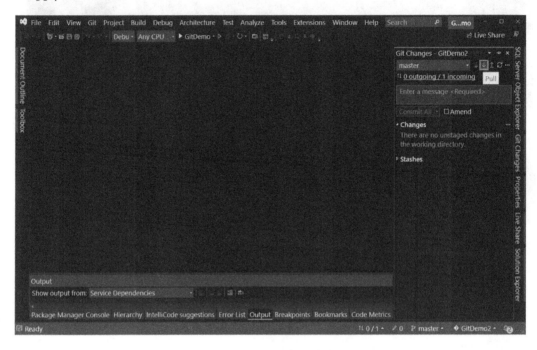

Figure 11.16 – The button for pulling changes

Once the changes have been downloaded, a message will appear, indicating which commit has been applied to the current project, as shown in *Figure 11.17*:

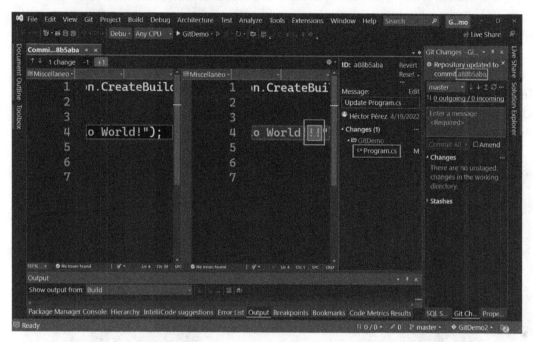

Figure 11.17 – The window showing the applied commit and the changes in it

If we click on the name of the commit, a window will appear, showing the changes that have been applied to the project, as shown in *Figure 11.17*.

In this section, although we have made a change in the remote repository from the main GitHub site, the most common way to do it is from Visual Studio itself. That is why, in the next section, we will analyze how to perform push operations to the repositories.

Pushing to repositories

A push operation refers to uploading changes to a repository. To demonstrate how this operation works, let's open the `Program.cs` file and modify line four, changing the `Hello World` string to `Hello Git`:

```
app.MapGet("/", () => "Hello Git!!!");
```

Immediately, you will see that the file icon in the **Solution Explorer** changes to a red checkmark, indicating that there has been a change in the local repository, which we can upload to the repository on GitHub:

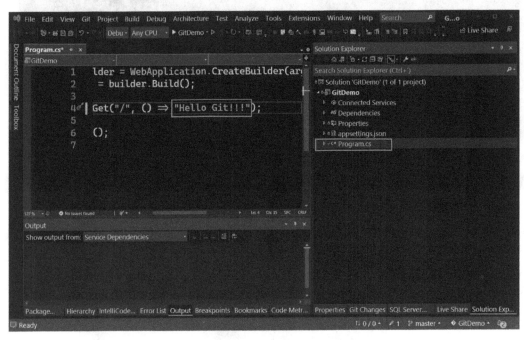

Figure 11.18 – The checkmark indicating that the original file has been modified

Similarly, in the **Git Changes** window, you will see a list of those files that have changes and can be uploaded to the repository, as shown in *Figure 11.19*:

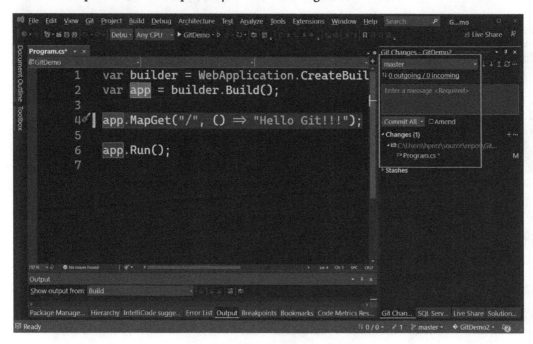

Figure 11.19 – The list of files with changes that can be uploaded to the remote repository

In the preceding figure, we can also see a button with the **Commit All** legend. This button is used to create a commit locally, without affecting the remote repository. This is a drop-down button that contains more options we can use, such as the **Commit All and Push** option. This option will push the changes to the remote repository.

For our demonstration, we will add the **Modified Program.cs** message, as shown in *Figure 11.20*, and click on the **Commit All and Push** option, to apply the changes on the server:

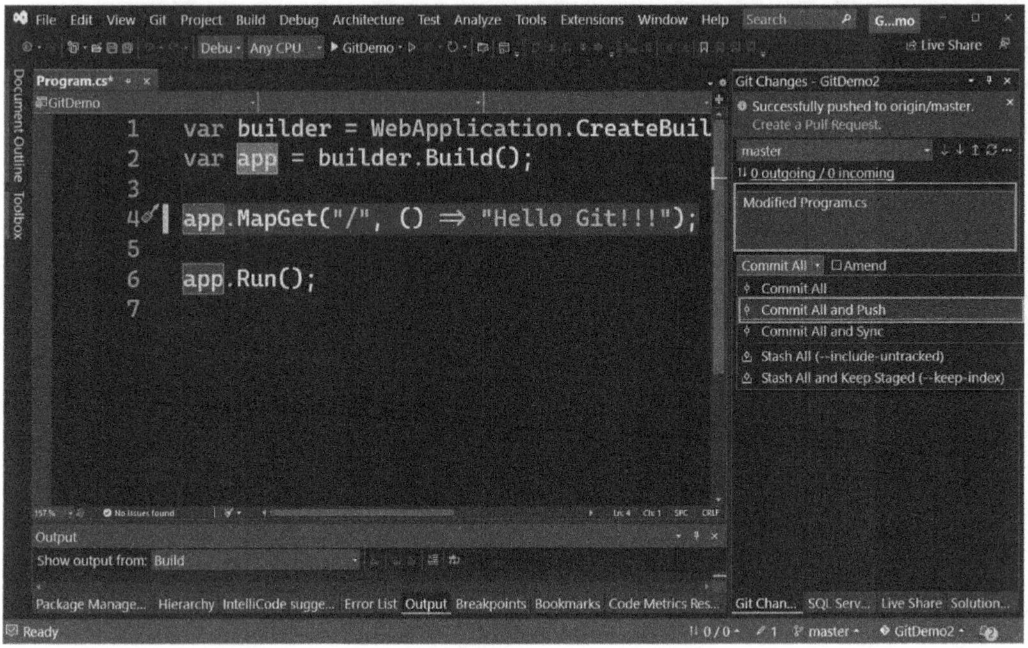

Figure 11.20 – The message added and the Commit All and Push option

If you go to the online repository in your GitHub account, you will see that the changes have been applied successfully.

> **Important Note**
>
> It is recommended that before executing a push operation, you always perform a pull operation to avoid merge conflicts as much as possible. You can use the **Sync (Pull then Push)** button to perform both actions one after another.

Now that we have analyzed how to execute the most common operations in GitHub, let's see how to work with branches in Visual Studio.

Managing branches

So far, we have been working with the main branch of our project called **master**. Imagine this branch as a timeline, where each event is performed by a commit. This is very useful when there is some conflict and you need to go back to a previous version, undoing the changes of a specific commit.

However, if you are working in a team, it is common that you will need to add a functionality in some kind of sandbox, before merging this functionality into the master branch. It is in this sort of scenario where Git branches will help us, allowing us to create a new project branch from an existing repository branch and work on it without affecting the functionality of the main repository.

To create a new branch, just go to the **Git | New Branch** menu. This will open a new window that asks for the branch name, the branch on which the new branch will be based, and a checkbox labeled **Checkout branch**, which, if checked, will transition to the new branch once it is created.

For this demonstration, let's use branch01 as the name of the branch, which will be based on the **master** branch, and leave the checkbox selected, as shown in *Figure 11.21*:

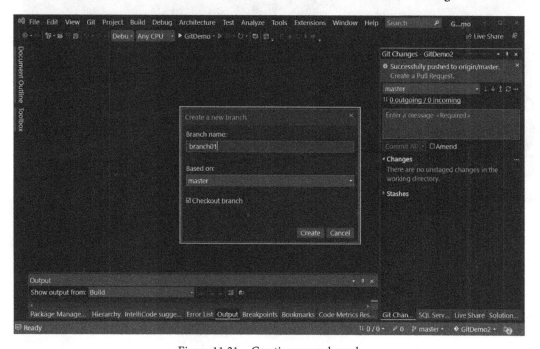

Figure 11.21 – Creating a new branch

Once the new branch has been created, we can apply as many changes as we need. In our example, we will open the Program.cs file and modify line four again, with the following code:

```
app.MapGet("/", () => "Hello Git!!! - branch01");
```

Once this change has been made, we will push the code as shown in the *Pushing to repositories* section, making sure to apply the change to **branch01**, as shown in *Figure 11.22*:

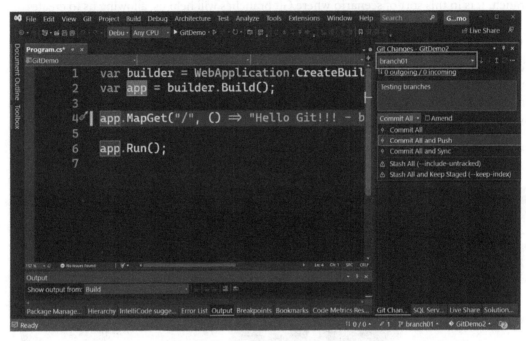

Figure 11.22 – Pushing changes to the new branch

This allows us to add functionality on an isolated remote repository without affecting the operation of the **master** branch.

Once you have your code tested, you will probably want to integrate it back into the **master** branch. This can be done from the **View | Git Repository** window. In this window, we will have a section called **Branches**, where we will see the list of the different branches in our project. Just right-click on the **master** branch and click on the **Checkout** option to switch to the master branch, which is where we will add the **branch01** changes. Then, right-click on the **branch01** branch and select the **Merge 'branch01' into 'master'** option:

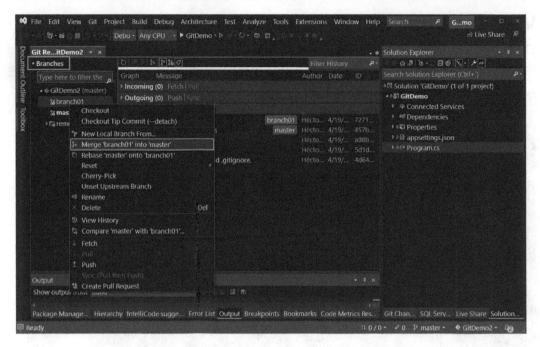

Figure 11.23 – Merging branches

This will cause the **branch01** branch to be merged with the **master** branch, which means that the new functionality will be integrated into the project.

Now that we have seen how to create branches and merge them, it is time to see how Visual Studio allows us to visualize the changes while editing the source code.

Viewing changes in repositories

There are several ways in which Visual Studio helps us to visualize the changes in the repositories. The first one is through the **Git Repository** window, as shown here:

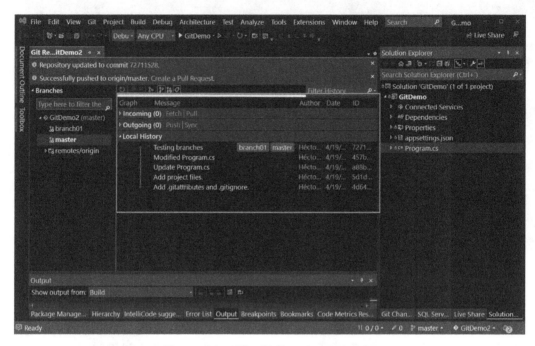

Figure 11.24 – The Git Repository window

This window allows us to visualize whether there are versions that have not been applied in our local repository through **Incoming**, whether there are commits made locally but not pushed to the server through **Outgoing**, and also **Local History**.

If we want to visualize the changes that have occurred between different commits, just right-click on two or more commits and select the **Compare Commits…** option:

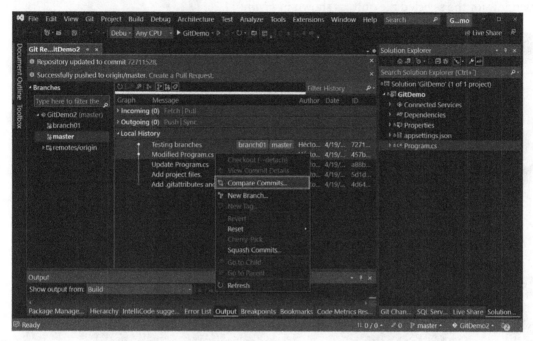

Figure 11.25 – The Compare Commits… option

This will display a new window with the changes that occurred between the different source files.

Another way to view changes to a single file is to right-click on a file in the **Solution Explorer** and then select the **Git | View History** option, which will open the changes window for the selected file only, as shown in *Figure 11.26*:

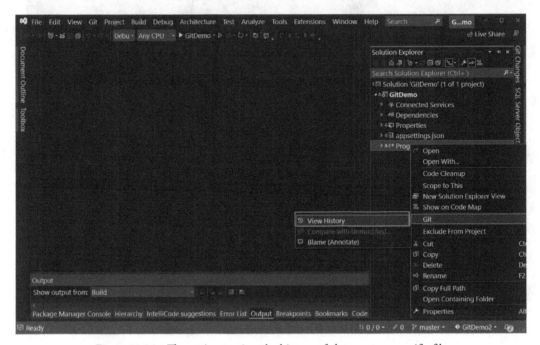

Figure 11.26 – The option to view the history of changes to a specific file

Finally, we already discussed the CodeLens functionality in *Chapter 7*, *Coding Efficiently with AI and Code Views*, which also contains functionality to view changes to the members of a class, such as the history of changes and who made modifications to a member.

Summary

In this chapter, we have learned how Visual Studio integrates tools so that we can easily manage projects using Git and GitHub.

Knowing how to work with Git-based projects is indispensable for all developers who want to manage their projects in a more controlled way, with the purpose of having an overview of a project structure in previous versions.

Likewise, if you work with other developers, you will be able to share project tasks, test them independently, and merge them when the code is reliable, in order not to damage the source code that has already been tested.

That is why we have learned how to set up a GitHub account in Visual Studio, how to create and clone repositories, how to perform fetch, pull, and push operations, how to manage branches in our projects, and finally, how to visualize changes in repositories.

In the next chapter, *Chapter 12, Sharing Code with Live Share*, you will learn how to work collaboratively with a development team on the same project in real time, thanks to the use of Live Share.

12
Sharing Code with Live Share

Collaboration tools are a new way to work remotely and make it easier for a group to achieve one common goal. It's amazing to see how we can edit the same document, record, or share resources in real time with other users. If we want to collaborate on coding or perform **pair programming**, which is a technique where two programmers work together on the same code, we need to be working in the same place, on the same machine, or use a tool to use a video call to perform these activities.

Visual Studio (VS) Live Share or just Live Share is a new tool included in VS by default that helps us to share our code with other programmers that use VS and VS Code.

We will review the following topics in this chapter:

- Understanding VS Live Share
- Using Live Share
- Performing live editing
- Sharing a terminal with other collaborators

First, we need to review Live Share in general and know how to find it.

Technical requirements

To use VS Live Share in VS 2022, you must have previously installed VS 2022 with the web development workload, as shown in *Chapter 1, Getting Started with Visual Studio 2022*. It's also important to have the SPA base project created in *Chapter 4, Creating Projects and Templates*.

You can check the changes made to the project in this chapter at the following link: `https://github.com/PacktPublishing/Hands-On-Visual-Studio-2022/tree/main/Chapter12`.

Understanding VS Live Share

VS Live Share is a real-time collaboration tool for programmers that use VS and VS Code. Live Share was launched as an extension for VS 2017, with some trial features available at that time. In VS 2022, it is included by default and contains all the features.

Live Share is completely free and can also be used in VS Code by installing an extension that you can find at this link: `https://marketplace.visualstudio.com/items?itemName=MS-vsliveshare.vsliveshare`.

Live Share has tools to edit, debug, share terminals, and execute projects for remote developers. We don't need to clone the repository or install additional extensions to see code and interact with it.

To complement the information provided during this chapter and the functionalities that you will review in the next section, you can read the documentation at this URL: `https://docs.microsoft.com/visualstudio/liveshare/`.

These kinds of collaboration tools are not new. Other IDEs have extensions and components to share code and work in real time. You can try the following tools and compare them with Live Share:

- **Teletype for Atom** (`https://teletype.atom.io/`): A collaboration tool to create shared workspaces between developers

- **Duckly** (https://duckly.com/): A real-time collaboration tool with video calls and other features compatible with many IDEs

- **CodeTogether** (https://www.codetogether.com/): A tool to create shared coding sessions that supports VS Code, IntelliJ IDEA, and Eclipse

> **Important Note**
>
> There are no other free collaboration tools for VS. Live Share is supported by Microsoft and the community.

In this section, we provided a brief overview about VS Live share, its history, and alternative tools that perform similar tasks. Now, let's start to use Live Share.

Using Live Share

To start using Live Share, we can go to the icon located at the top right in the main window in VS, as shown in *Figure 12.1*:

Figure 12.1 – The Live Share button in the main window

After clicking on this icon, you will see a new window, where you can select the account to use to create a new live share session. This account is required, and the account used in VS is selected by default. Also, we can see a **Sharing…** message where the **Live Share** button is located (see *Figure 12.2*):

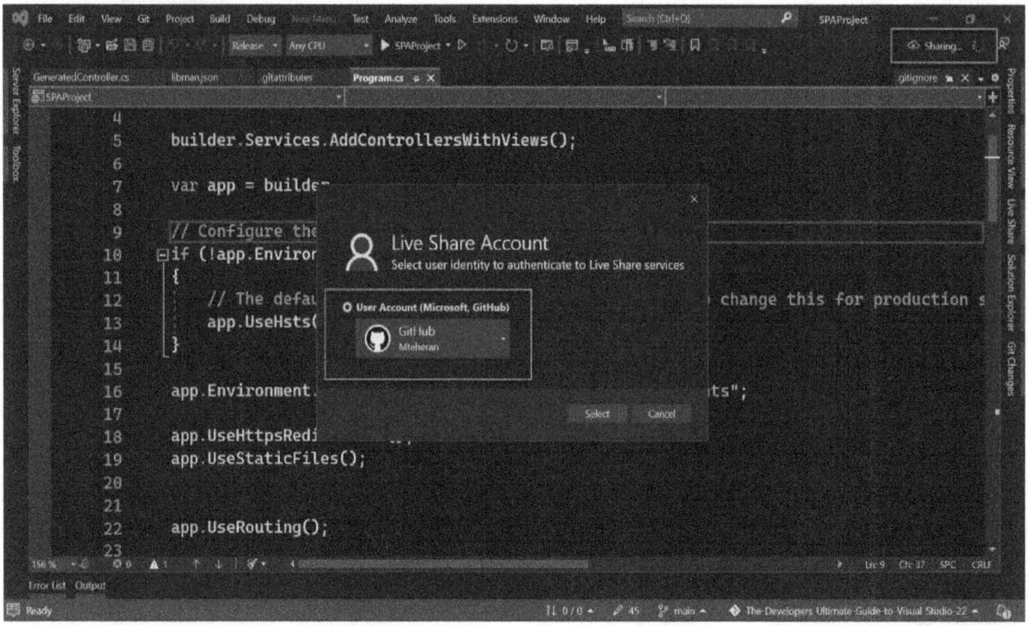

Figure 12.2 – The Live share account window

You can select the account that you want to use and then click on **Select**. Then, you will get a confirmation message, and the link to share the session is automatically added to the clipboard. The live share button displays **Sharing…**, which confirms that the session is live (see *Figure 12.3*):

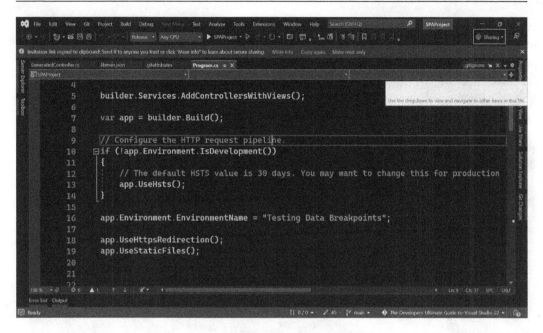

Figure 12.3 – VS with a live session in Live Share

You can share the invitation link with other developers, coworkers, or friends to try this tool. This is an example of a link generated by VS Live Share: `https://prod.liveshare.vsengsaas.visualstudio.com/join?7E72234CE1703CF92015D01564C560706AE1`.

When someone opens the link in a browser, they are going to see **Visual Code for the web**. This is a version of the VS Code editor that runs as a web page with no dependencies on the operating system or additional requirements. To read more information, visit this link: `https://code.visualstudio.com/docs/editor/vscode-web`. In VS Code for the web, you have the option to continue as anonymous or sign in using a GitHub account:

Figure 12.4 – A session opened in the browser

We can use **Continue as anonymous** to continue and join the session. Three options will be shown that we can choose to edit and navigate in the code (see *Figure 12.5*):

Figure 12.5 – Options to open the session with the link

Let's review these three options further:

- **Continue in Web**: With this option, you will continue using VS Code in the browser.

- **Open in Visual Studio Code**: This option will open VS Code on your computer and you will be able to edit the code there.

- **Open in Visual Studio**: This option will execute VS on your local machine and you will be able to edit the code there.

You can use **Continue in Web** to join in a session using the browser. The user that created the session will see a new notification message, where you can see the collaborators that are trying to join the session. You can accept the new collaborator and continue with the session (see *Figure 12.6*):

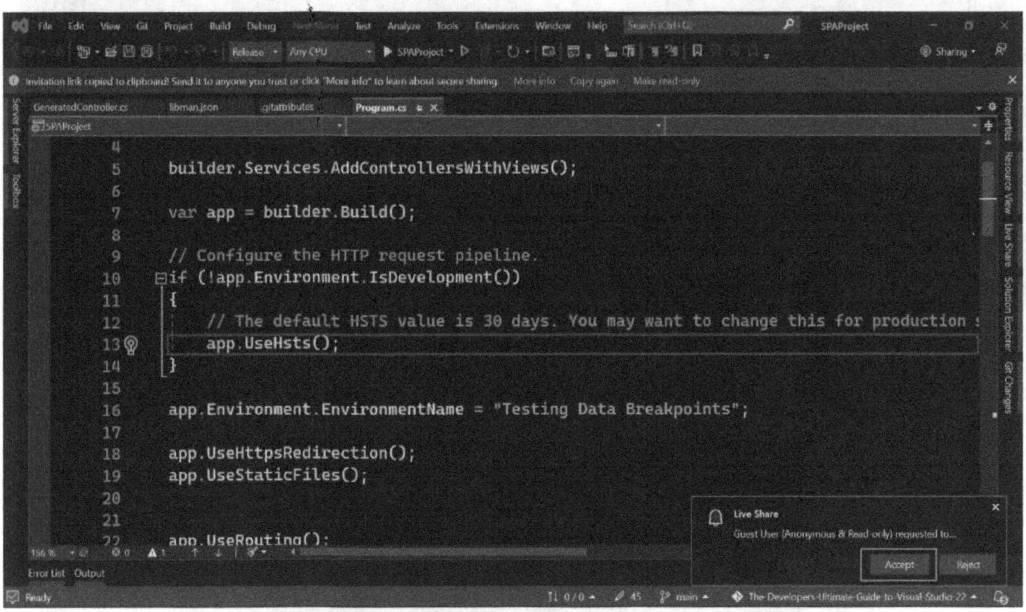

Figure 12.6 – A notification from Live Share to accept a new collaborator

After accepting the guest user, Live Share will share the code and show the changes in real time. In this case, the guest user can see the moves in code performed by *Miguel Teheran*, who created the session. See the **Following Miguel Teheran** message in yellow in *Figure 12.7*, which also shows the option to open the session in VS Code and VS:

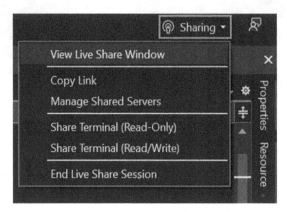

Figure 12.7 – The option to open the session in VS

In VS, there is a menu with different options that we can use during the session:

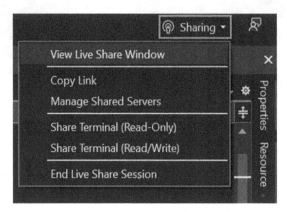

Figure 12.8 – The Live Share menu during a live session

Let's review these options and functionalities further:

- **View Live Share Window**: View the session status, including the session details and collaborators in the session.

- **Copy Link**: Copy the link in the clipboard to share it with others.

- **Manage Shared Servers**: Open a new window where we can share our local server with other users in the session.

- **Share Terminal (Read-Only)**: Share the terminal in read-only mode to share the console log and the results with others. Collaborators cannot execute any commands in the terminal.

- **Share Terminal (Read/Write)**: Share the terminal with others in the session, with the possibility to execute commands remotely.

- **End Live Share session**: End the session for all the users connected.

Let's click **View Live Share Window** to see the session status. It will show a new window in the right panel, as we can see in *Figure 12.9*:

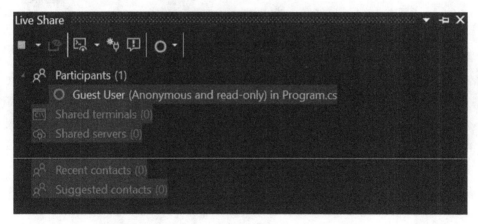

Figure 12.9 – Live Share windows during a session

In the **Participants** section, we can see the guest user who is watching the code in `Program.cs`. We can also see the shared terminals and servers. With the red square, we can end the session at any time for all the participants.

> **Important Note**
>
> Guest users join the session in read-only mode, which means that they cannot modify or update the code. This is a security requirement. To read more about security in Live Share, visit `https://docs.microsoft.com/en-us/visualstudio/liveshare/reference/security`.

It's quite simple to create and share a session using Live Share. We can see the collaborators connect and end the session at any time. Let's see how to edit or update the code in real time.

Performing live editing

We can share a Live Share session with others, but those users can only join in either edit or read-only mode. By default, all guest users are in read-only mode, so let's join the session with VS using a logged user.

After selecting **Open in Visual Studio**, as shown in *Figure 12.5*, VS will load the project and show a loading message to finally show the code with the active session. We will have a menu where the live share button is located to manage the session as a collaborator (see *Figure 12.10*):

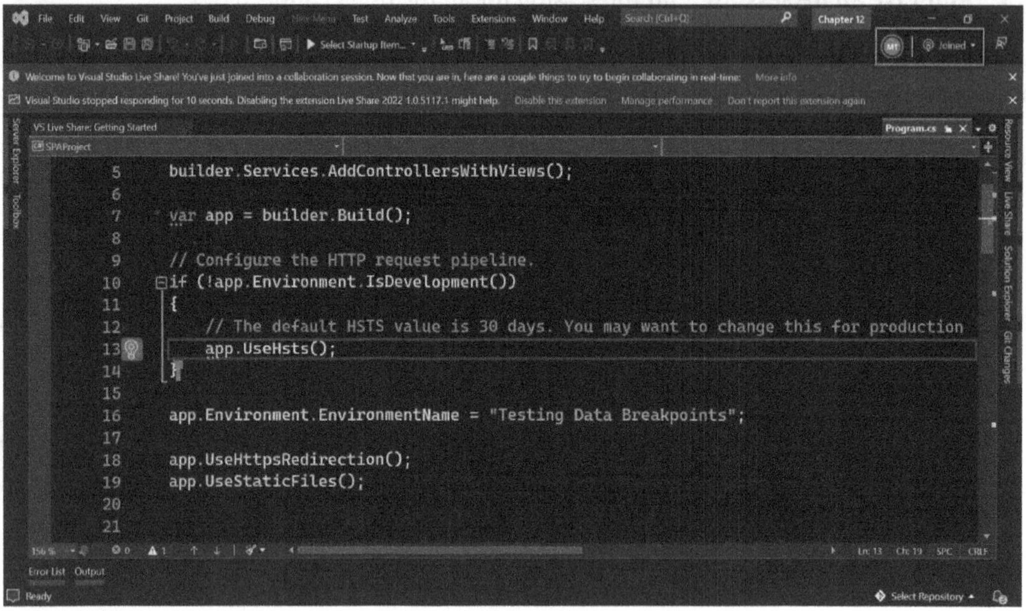

Figure 12.10 – VS when a collaborator joins in a session

In the preceding figure, we can also see some icons on the left, representing the active collaborators in the session. We can click on **Joined** to see the options that we have as a collaborator in an active session (see *Figure 12.11*):

Figure 12.11 – The Live Share menu for collaborators

Let's review each option in this menu:

- **View Live Share Window**: This option allows you to view the session status, including the session details and collaborators in the session.

- **Session Chat**: This opens a new window where you can type and share messages with other collaborators.

- **View Shared Servers**: With this option, you can visualize the shared servers in the current session.

- **Show Shared Terminals**: This option opens a new window where you can see the currently shared terminals.

- **Focus Participants**: With this option, you will keep the focus on what other collaborators are doing or editing.

- **Leave Live Share Session**: With this option, you will leave the current session, while other collaborators can continue in the session.

> **Important Note**
>
> The host is the only participant that can end the session. All the collaborators in the session will receive a notification when the session ends. For more information, visit `https://docs.microsoft.com/visualstudio/liveshare/use/share-project-join-session-visual-studio`.

Now, you can edit any part of the code and see how Live Share works. For example, in line seven of the `Program.cs` file, we can add a comment for the `build` method (see *Figure 12.12*):

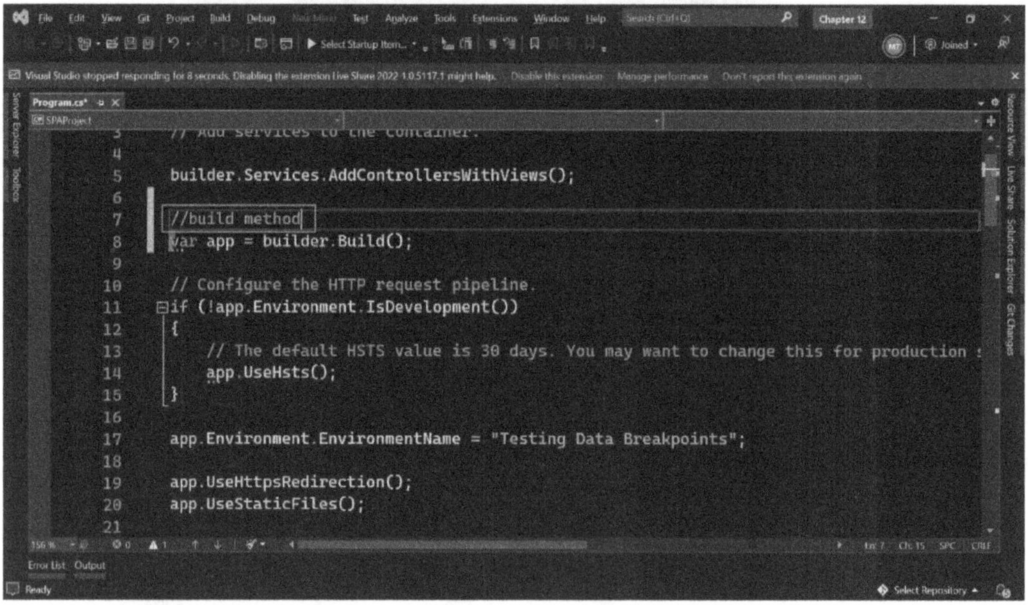

Figure 12.12 – Adding a comment during a Live Share session

Once the collaborator starts typing, the host and other users in the session can see the changes in the file in real time. The collaborator's name will be displayed in the specific line, highlighted with a random color (see *Figure 12.13*):

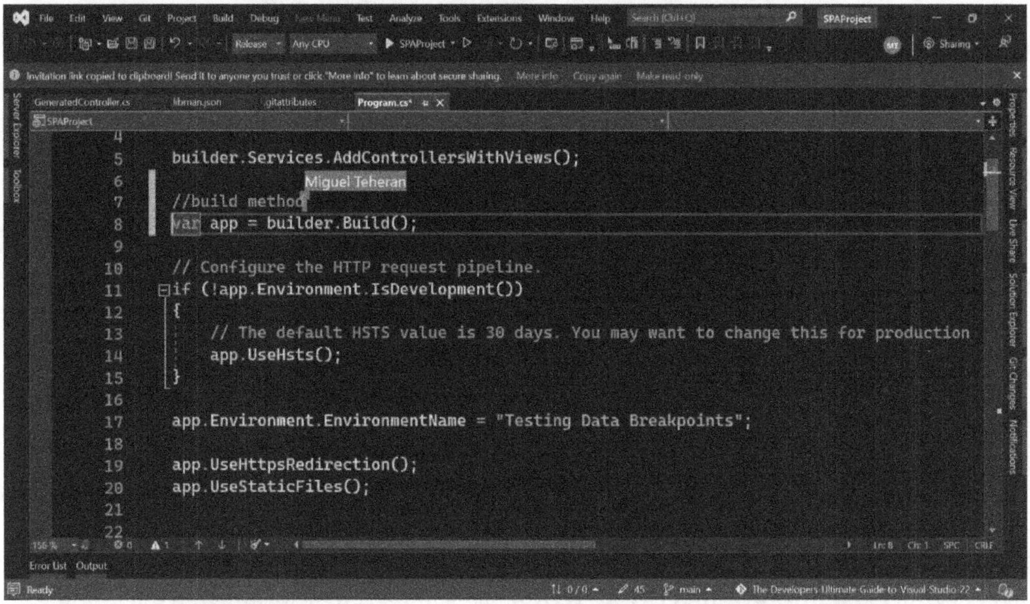

Figure 12.13 – The host watching changes in real time made by a collaborator

> **Important Note**
> Live Share will assign a color for each new collaborator randomly to easily identify each user during the session.

You can edit the file to perform a suggestion, but you can also save the file using the **File | Save Selected Items** menu or the *Ctrl + S* shortcut.

Now that we know how Live Share works and some of the features that we can use during a shared session, we are ready to review an option to share a terminal with other collaborators during a session.

Sharing a terminal with other collaborators

In Live Share, we can also share a terminal with other developers. By sharing the terminal, we allow other developers to see more details of a project using the command line. After creating a new session, we can use the **Share Terminal** options to allow others to use our terminal (see *Figure 12.14*):

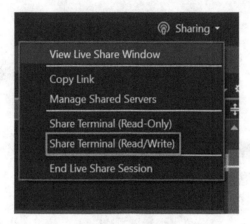

Figure 12.14 – The Share Terminal options in the Live Share menu

After clicking on **Share Terminal (Read/Write)**, we will see a new terminal in a window with a mark that indicates that the terminal is shared. In *Figure 12.15*, we can see the new terminal added in VS in the bottom panel:

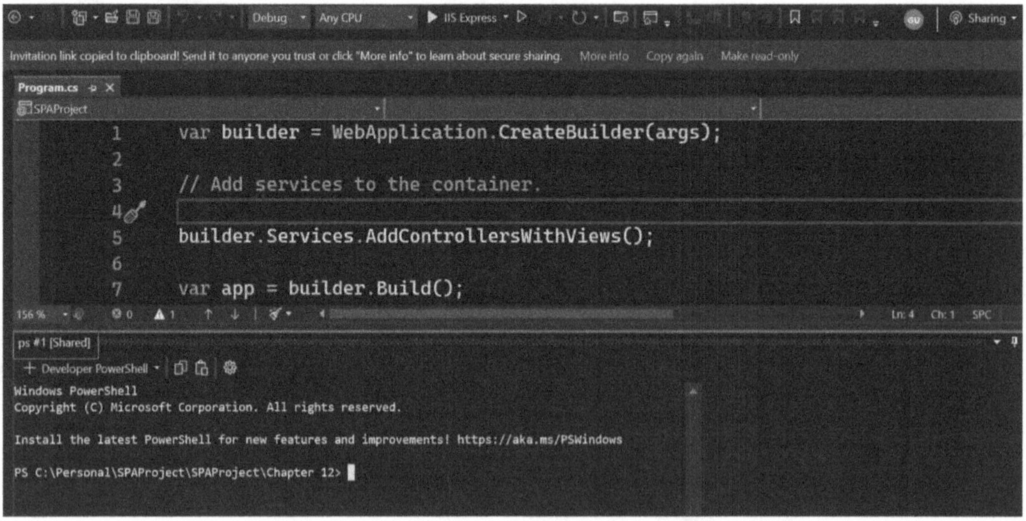

Figure 12.15 – A terminal shared during a Live Share session

Other collaborators will see this shared terminal automatically when they join the session (see *Figure 12.16*):

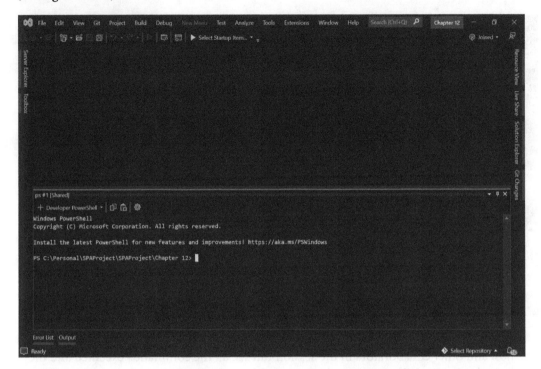

Figure 12.16 – A terminal shared during a Live Share session

The collaborator in the session can execute any command in the project to get more details or add information using the windows commands. In this case, we can use the .NET **command-line interface (CLI)**, which is a tool included in .NET, where we can compile, restore, run, and deploy .NET-based projects using simple commands in the terminal. To interact with the project, you can compile it using the `dotnet build` command. To learn more about the .NET CLI, you can check out the documentation at the following link: `https://docs.microsoft.com/dotnet/core/tools/`.

We will see warnings and errors during the compilation process in the terminal (see *Figure 12.17*):

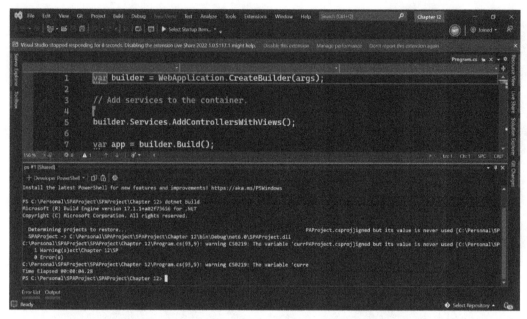

Figure 12.17 – A collaborator executes the dotnet build command in the shared terminal

The host and other collaborators in the session can see all the commands executed and the results in the terminal. The command is executed in the host's environment, which means the user that created the session (see *Figure 12.18*):

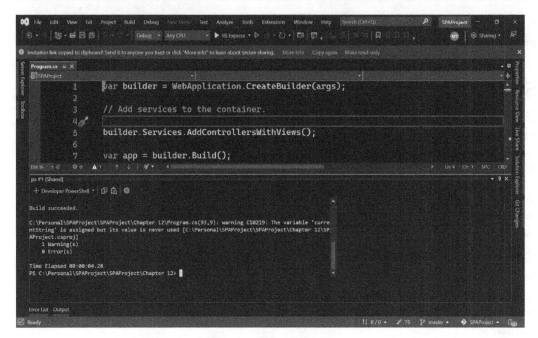

Figure 12.18 – The host of the session watches the results of the dotnet build command

If the terminal is not displayed by default or you want to see which other terminals are shared in the session, you can go to the **Live Share** window and see the status of the current session, including the shared terminals and the permissions for them (see *Figure 12.19*):

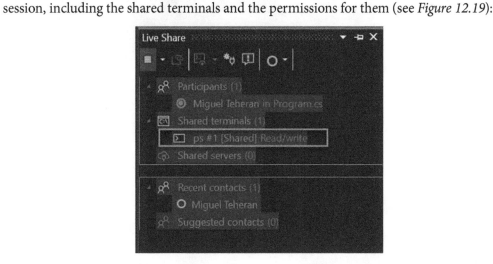

Figure 12.19 – The Live Share window with a shared terminal

Now that you know how to use Live Share in VS 2022 as a host and collaborator, you can invite others to collaborate on your project and share a terminal when it's required.

Summary

Live Share is an amazing tool to collaborate with others in real time. With globalization, this tool is now more important than ever, considering that a global development team can include developers working around the world in different time zones and with different tools.

You have learned how to use Live Share to work as a team and collaborate with other developers on the same project. You know how to load a Live Share session in VS, see the participants in the session, share a terminal, and end the session whenever you want. Also, you reviewed some alternatives to Live Share with other IDEs and editors.

In *Chapter 13*, *Working with Extensions in Visual Studio*, you will study how to increase the tools and functionalities included by default in VS using extensions. You will learn how to search, install, and set up extensions in VS and how to take advantage of them to increase your productivity.

13
Working with Extensions in Visual Studio

We cannot deny that Visual Studio's native functionality for performing tasks, as we have seen so far, is phenomenal. However, there may be times when you want to extend the capabilities of the IDE with simple features, such as applying a new theme with custom colors, or complex functionality, such as code-refactoring tools.

It is in these cases where Visual Studio extensions are of great help, which is why we will dedicate an entire chapter to an analysis of their use.

The main topics we will see are as follows:

- Working with the extensions tool
- Searching for and installing extensions
- Reviewing Visual Studio Marketplace
- Setting up extensions
- Creating a new theme as an extension

Let's start our tour through the extensibility of Visual Studio, thanks to extensions.

Technical requirements

To follow this chapter, you must have installed Visual Studio with the web development workload, as shown in *Chapter 1, Getting Started with Visual Studio 2022*.

Additionally, to create a theme as an extension, as shown in the *Creating a new theme as an extension* section, you must install the **Visual Studio extension development** workload, as shown in *Figure 13.1*:

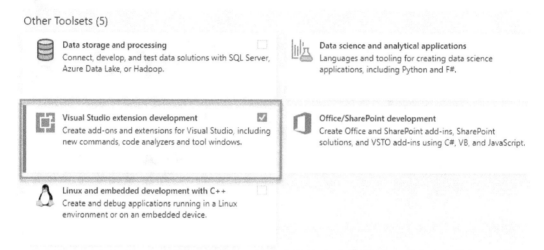

Figure 13.1 – Installing the Visual Studio extension development workload

You can install it before you start reading the *Creating a new theme as an extension* section or while creating the extension, as shown in the same section.

As the chapter focuses on showing the use of extensions in Visual Studio, a code repository is not required.

Let's see how to work with the extensions tool from Visual Studio.

Working with the extensions tool

The main purpose of extensions in Visual Studio is that you can improve your day-to-day productivity with features that may be somewhat specific to particular tasks, such as providing suggestions for best practices in code, performing code cleanup, highlighting messages in the output window, adding visual features in code files, or interacting with SQLite databases. Allowing developers to create new features for Visual Studio and share them with the rest of the world is an excellent move on the part of the Microsoft team.

> **Important Note**
>
> In this section, I will show you simple extensions to familiarize you with the concept of extensions. In *Chapter 14, Using Popular Extensions*, you will learn about the most widely used and preferred extensions by developers and how to work with them.

But how can we access these extensions? The most direct answer is through the extensions tool, which you can find through the **Extensions | Manage Extensions** menu, as shown in *Figure 13.2*:

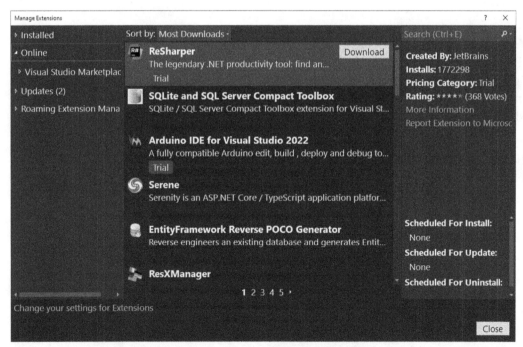

Figure 13.2 – The extensions tool

On the left side, we find the following categories, which allow us to filter the extensions:

- **Installed**: These are the extensions that we have previously installed in the Visual Studio instance.

- **Online**: In this category, you will find the extensions available to be added to Visual Studio. In the upper part, we have a filter with the **Sort by** legend, which allows us to filter the extensions in different ways.

- **Updates**: This category shows the extensions that have pending updates, either with bug fixes or extension improvements.

- **Roaming Extensions Manager**: This category allows you to view extensions previously installed in an instance of Visual Studio that are associated with a Microsoft account, so they can be installed in the current instance of Visual Studio. This means that if you are logged in on a different computer and you install an extension, you will be able to see it in this list, so you do not have to remember the name of the extension and can install it easily.

Now, let's see how we can search and install extensions, thanks to this tool.

Searching for and installing extensions

The extensions tool, which is shown in *Figure 13.3*, has a search box located at the top right, where we can enter a search term to find extensions referring to some technology or tool. It is important to note that the search will be performed in the selected category, which we discussed in the *Working with the extensions tools* section.

If you want to perform a search among all the extensions, the best thing to do is to go to the **Online** category and perform the search. In our example, let's search for extensions related to the keyword `javascript`:

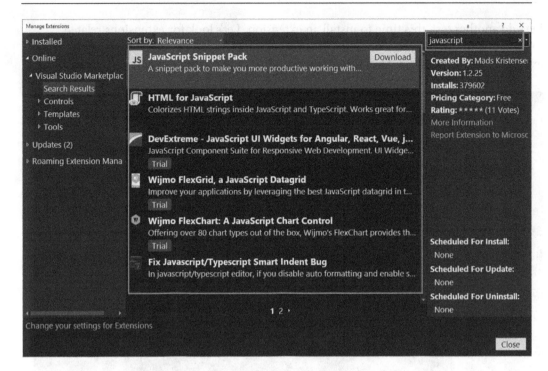

Figure 13.3 – A list of results in the extensions tool

The order of the list of extensions will appear according to the drop-down item at the top, which by default appears as sorting by **Relevance**, but we can also select other options, such as displaying by **Most Downloads**, **Most Recent**, **Trending**, or **Highest Rating**.

Let's do another test, this time to install an extension. Let's look for extensions related to css, and in the list, you will see an extension called **Color Preview**. To install an extension, just click on the **Download** button, which will start the download process:

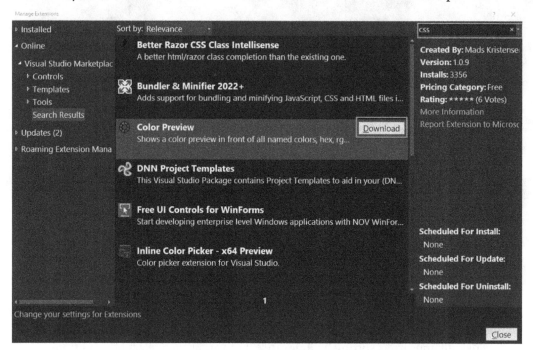

Figure 13.4 – Searching for and installing an extension

Once the extension download is complete, a message will appear at the bottom of the window, stating **Your changes will be scheduled. The modifications will begin when all Microsoft Visual Studio windows are closed**. The message is very descriptive, so we will proceed to restart Visual Studio to perform the installation.

Once we close Visual Studio, the installation process will start, which will show you (as shown in *Figure 13.5*) the extension that will be installed on the Visual Studio instance, as well as additional information, such as the **Digital Signature** and **License** types:

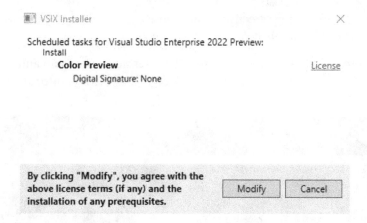

Figure 13.5 – The extension installation window

If we agree with the displayed information, as shown in *Figure 13.5*, we must click on the **Modify** button, which will start the installation process in the IDE.

Once the installation process is complete, you must reopen an instance of Visual Studio. Although at first glance it seems that nothing has changed, this extension has added the ability to show us the selected color applied in the properties of a `css` file to the IDE, as demonstrated in *Figure 13.6*:

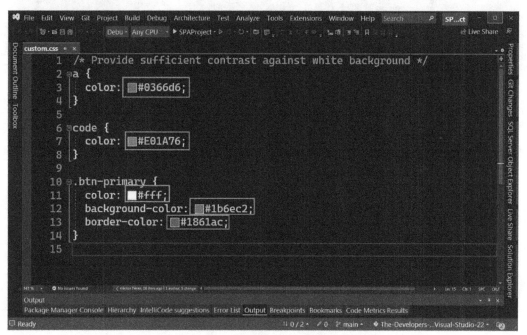

Figure 13.6 – The feature added by the Color Preview extension

You can compare the preceding screenshot with those shown in the *Working with CSS styling tools* section of *Chapter 9, Styling and Cleanup Tools*.

If you want to uninstall an extension, just open the extensions tool again and look for the extension in the **Installed** section. Then, click on the **Uninstall** button, as shown in *Figure 13.7*:

Figure 13.7 – Uninstalling an extension

In the preceding screenshot, you can see a second button called **Disable**, which allows you to temporarily disable an extension until you decide to reactivate it.

Now that we have seen how to install extensions from the extensions tool, let's take a look at Visual Studio Marketplace.

Reviewing Visual Studio Marketplace

Visual Studio Marketplace is the online place to find and install extensions for Visual Studio 2022. In this marketplace, you can also find extensions for other products of the Visual Studio family, such as Visual Studio Code and Azure DevOps. You can access the marketplace through the site: `https://marketplace.visualstudio.com/vs`.

Once you enter the portal, you will have a very different user interface to the extensions tool, but the core operation is the same. In the main portal, you will be able to see the different extensions sorted by **Featured**, **Trending**, **Most Popular**, and **Highest Rated**.

To test the marketplace, let's search for the term `icons` and see the list of results. Although the list currently yields 49 results, not all extensions are compatible with the most modern version of Visual Studio. Therefore, it is recommended to apply the filter for **Visual Studio 2022**, as shown in *Figure 13.8*:

Figure 13.8 – Filtering Visual Studio versions in the marketplace

Once the filter has been applied, find the **Visual Studio Iconizer** extension and click on it to go to the extension page:

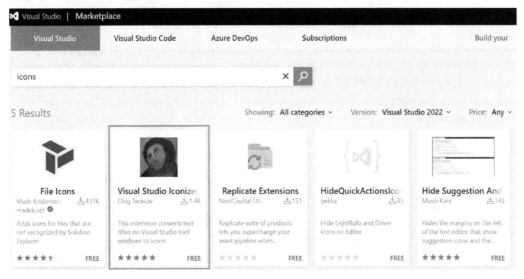

Figure 13.9 – Searching for extensions in the marketplace

On the extension page, you will be able to find information that will give you a general idea about the extension, such as the number of installations, the number of reviews, ratings, the change log, and the project page, among other data. If the description of the extension describes that it can help you solve a problem, you can click on the green **Download** button, as shown next:

Figure 13.10 – The button to download an extension

This will start the process of downloading a file with a `vsix` extension – in this particular case, the name is `VSIconizer.vsix`. To install the extension in Visual Studio, you must first make sure you have closed all instances of the IDE and then run the downloaded file, which will start the same installation process that we saw in the *Searching for and installing extensions* section.

Once the installation process is finished and an instance of Visual Studio 2022 is opened, we can see how the appearance of the IDE tabs has changed, illustrated in *Figure 13.11*, with the text being replaced by icons:

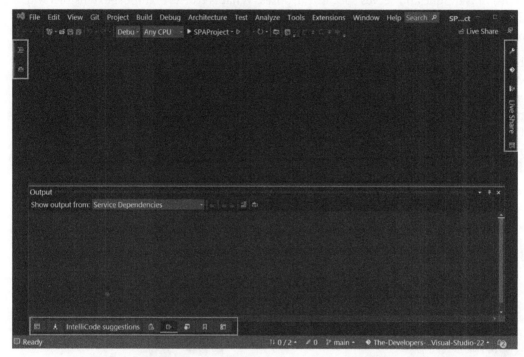

Figure 13.11 – The appearance of the IDE tabs modified by the Iconizer extension

Whether you opt for an installation with the extensions tool or by downloading the extensions from the marketplace, the installation process is extremely simple.

Now, let's see how to configure the extensions we install.

Setting up extensions

Unfortunately, there is no standardized way to configure a Visual Studio extension, as each extension is unique and serves specific purposes. What is true is that most extensions will add configuration options for the extension, either through a special window, or from the configuration options.

The best way to know these configuration options is through the extension page itself. For example, in the case of the **Visual Studio Iconizer** extension, which we installed in the *Reviewing Visual Studio Marketplace* section, the initial behavior is to show only the icons on the tabs. The extension page tells us that this behavior can be changed to show the text of the tab next to the icon added by the extension. This can be done through the option that has been added through the **Tools | Options | Environment | Iconizer** menu, as shown in *Figure 13.12*:

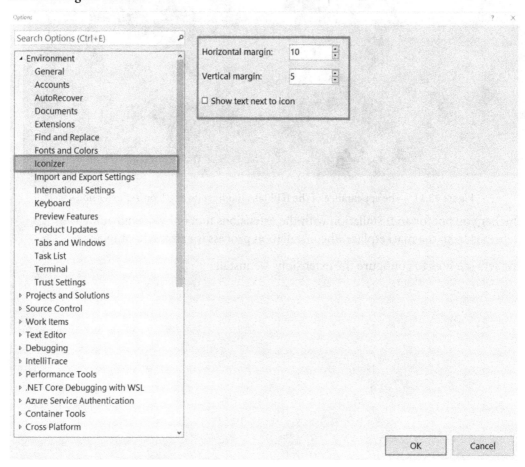

Figure 13.12 – The Iconizer extension options

Most extensions create a section, as shown in the preceding figure, and the more complex ones even provide a special new menu for the tool, which we will discuss in *Chapter 14, Using Popular Extensions.*

Now, let's look at a way to create a new theme for Visual Studio and export it as an extension so that more users can apply it to their Visual Studio instances.

Creating a new theme as an extension

If you want to carry out an even more customized color configuration for your IDE, it is possible to do so, thanks to the extension called **Visual Studio Color Theme Designer**. You can search and download this extension within Visual Studio Marketplace, as we saw in the *Reviewing Visual Studio Marketplace* section.

Once you have installed the extension, you will need to start Visual Studio and select the **Create a new project** option, as shown here:

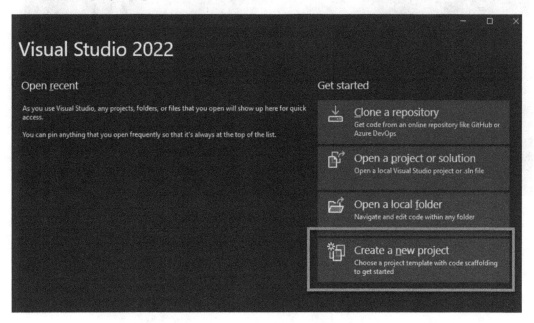

Figure 13.13 – Creating a new project

In the window that is opened, you must select the project from the list called **VSTheme Project**, which has been installed, thanks to the previously added extension. If you do not find it at first glance, you can search for it in the search box at the top, as shown in *Figure 13.14*:

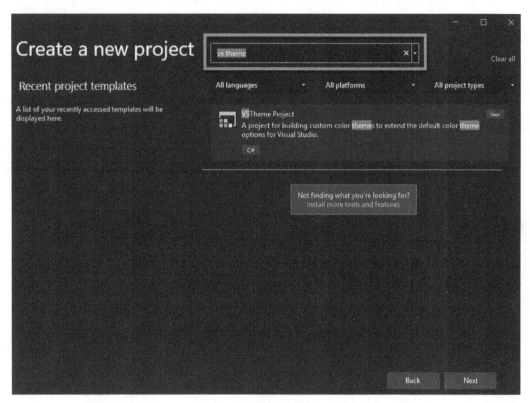

Figure 13.14 – Searching for the VSTheme project

In the new project configuration window, as shown in *Figure 13.15*, you can enter a project name that refers to the theme you are creating. For this example, I will name it CustomTheme. You can leave the other parameters as the default values. Proceed to create the new project by clicking on **Create**:

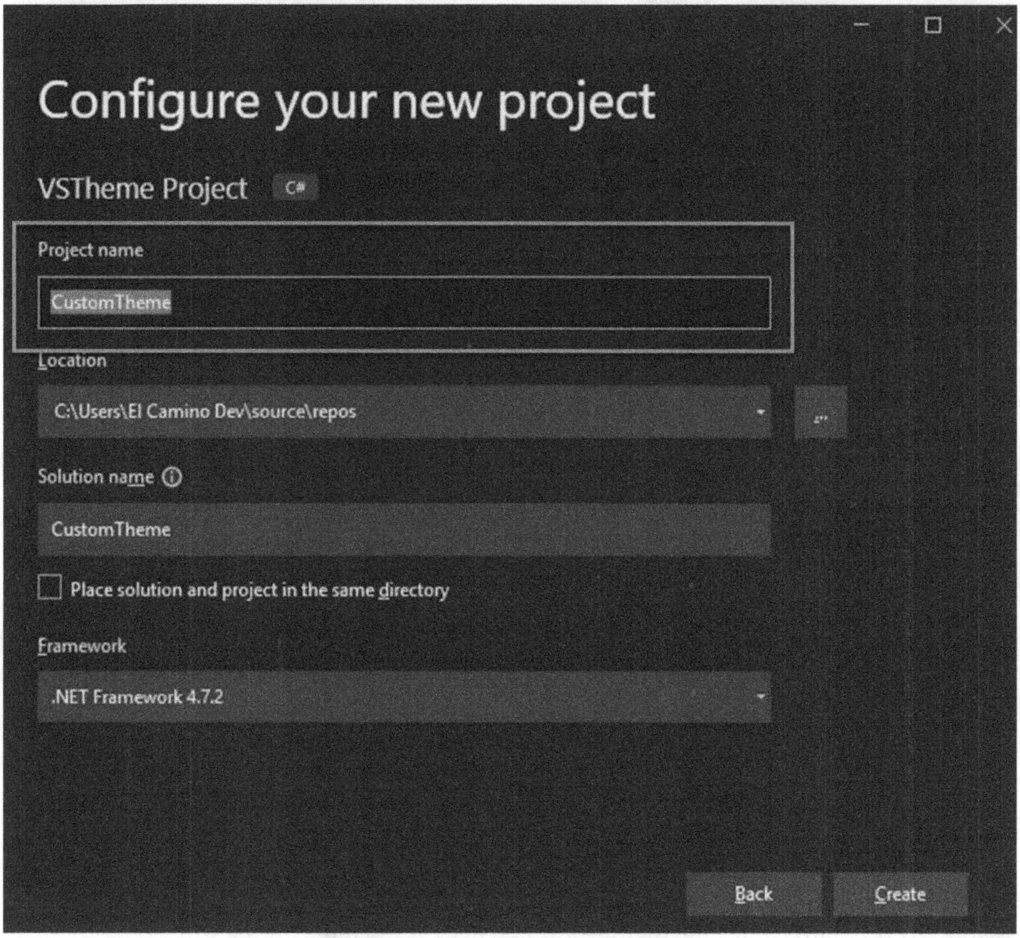

Figure 13.15 – Project configuration

If you are shown a window indicating that you must install the **Visual Studio extension development** workload, click on **Install**:

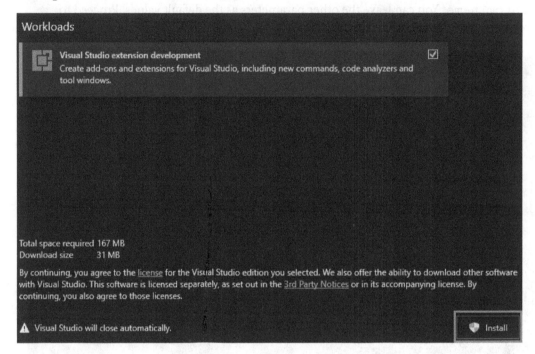

Figure 13.16 – Installing the Visual Studio extension development workload

This will enable the environment to be able to develop extensions for the IDE.

> **Important Note**
>
> If, after installing the workload shown in *Figure 13.16*, you get a solution without files, you must recreate the project, which will solve the problem.

Once the IDE has installed the extension, the new project will be created with different elements. To configure your new theme, double-click on the file with the `vstheme` extension, inside the **Solution Explorer** window, as shown here:

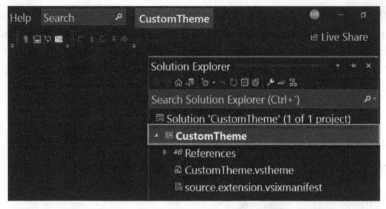

Figure 13.17 – Opening the theme configuration file

This will start a wizard, which will allow you to create your new theme quickly. You will start in a tab called **Quick start**, where you will be able to select a base theme from the installed ones in order to make modifications to the new theme:

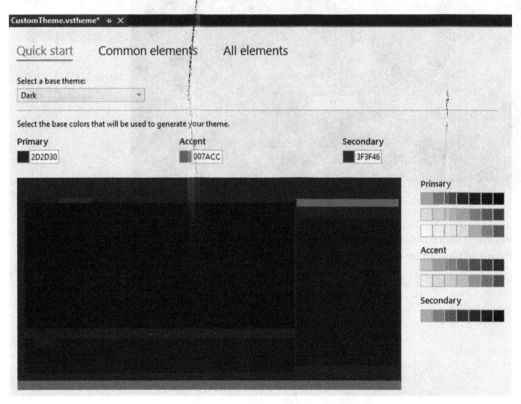

Figure 13.18 – Setting up a new theme

In *Figure 13.18*, the **Dark** theme has been selected for the demonstration. From here, you can click on any of the three main colors, tagged as **Primary**, **Accent**, and **Secondary**, to select a different color and apply it by clicking the **OK** button, highlighted in red in *Figure 13.19*:

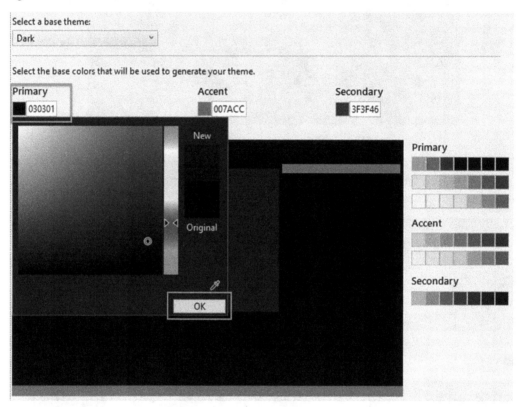

Figure 13.19 – Changing a color

If you want more detailed customization, you can do it by following the same procedure, both in the **Common elements** and **All elements** tabs, as shown in *Figure 13.18*.

Finally, to use the theme, you have a **Preview** button at the bottom of the window to preview how the changes would look with your new configuration, and a button with the legend **Apply** to apply the changes to the IDE:

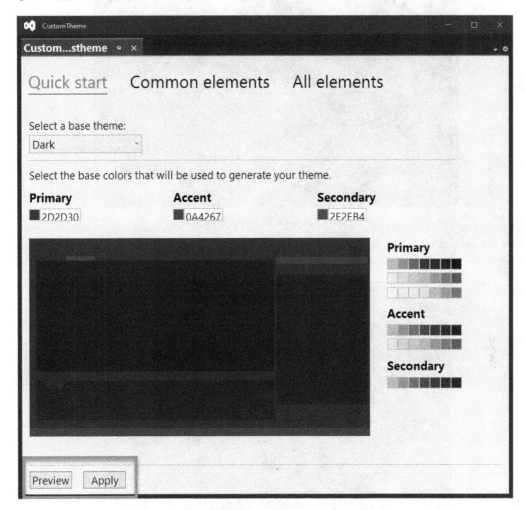

Figure 13.20 – The buttons to preview and apply theme changes

To obtain the installable extension file, you must compile the project. One of the ways to compile the project is shown in *Figure 13.21*:

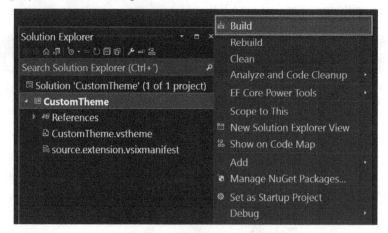

Figure 13.21 – Building the project

Once the project has been successfully compiled, we must go to the compilation directory by right-clicking on the project and selecting the **Open Folder in File Explorer** option:

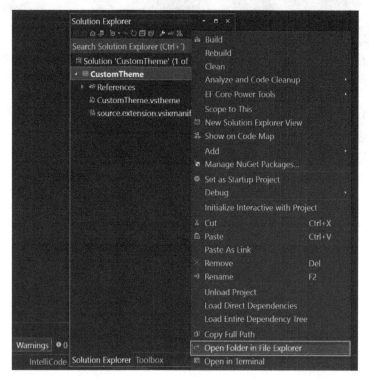

Figure 13.22 – The option to open the project folder in Windows Explorer

This will open the Windows file explorer at the project location. In this path, we will proceed to find the file located in the `bin | Debug | CustomTheme.vsix` path. The name of the file will be taken from the name of the project. In our example, the name of the project is `CustomTheme`, as shown in *Figure 13.15*, so if you have used another name for the creation of the project, you will find the file with that name. This is the file that you can distribute to share the theme, or you can upload it to Visual Studio Marketplace.

> **Important Note**
>
> It is also possible to set the output mode to **Release**, as discussed in *Chapter 5, Debugging and Compiling Your Projects*. This will generate a smaller file that will run faster.

As we have seen, creating a new theme, thanks to Visual Studio extensions, is extremely easy.

Summary

In this chapter, we have seen that extensions are a way in which we can extend the functionality of Visual Studio, always with the aim of improving our user experience and optimizing development time. We have tested a few extensions that have completely changed some of the IDE's functionality, such as previewing colors in `css` files, changing tabs to icons, and creating themes for distribution.

Also, we have reviewed how to search for and install extensions, both from the extensions tool and Visual Studio Marketplace. We analyzed how extensions are regularly configured, and finally, we created a new custom theme that can be shared with others, thanks to the use of an extension.

In *Chapter 14, Using Popular Extensions*, we will delve even deeper into the topic of extensions, analyzing which ones are the most popular because of their usefulness in web development.

14
Using Popular Extensions

In *Chapter 13*, *Working with Extensions in Visual Studio*, we learned how to extend utilities and functionalities by installing extensions published by the Visual Studio community and third-party vendors. We can install these extensions using the extension manager in Visual Studio and then restart Visual Studio to see the changes in the user interface.

In this chapter, we will analyze some free and useful extensions to increase productivity and improve our experience of using Visual Studio. We will install and review the extensions in the following sections:

- Adding HTML Snippet Pack
- Cleaning up code with CodeMaid
- Compiling web libraries with Web Compiler
- Identifying white space with Indent Guides

We will start with HTML Snippet Pack, an extension that helps us include some additional code snippets in Visual Studio when coding HTML files.

Technical requirements

To install the extensions in Visual Studio 2022, you must have previously installed Visual Studio 2022 with the web development workload, as shown in *Chapter 1, Getting Started with Visual Studio 2022*. It's also important to have the SPA project created in *Chapter 4, Creating Projects and Templates*.

You can check out the changes made to the project in this chapter at the following link: `https://github.com/PacktPublishing/Hands-On-Visual-Studio-2022/tree/main/Chapter14`.

Adding HTML Snippet Pack

In *Chapter 6, Adding Code Snippets*, we reviewed how code snippets can improve our productivity while we are coding. We also learned how to create, import, and remove code snippets using the Code Snippets Manager.

In Visual Studio's extension marketplace, we can find many extensions that add code snippets for different technologies by navigating to `https://marketplace.visualstudio.com/` and typing `snippet` in the search bar (see *Figure 14.1*):

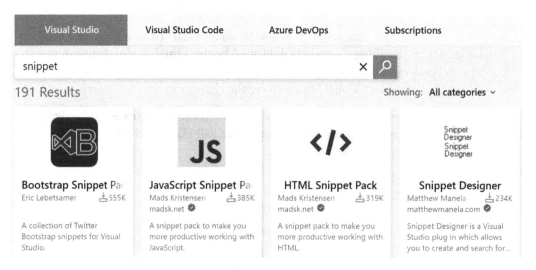

Figure 14.1 – Extensions related to snippets in the Visual Studio Marketplace

One of the most popular snippet packs for web developers is **HTML Snippet Pack**. With this extension, you can code in HTML faster, creating pieces of code and HTML elements after typing some charactes in the editor. Let's install this extension and see how we can create HTML elements easily using snippets.

Downloading and installing HTML Snippet Pack

Navigate in Visual Studio to **Extensions | Manage Extensions** and type HTML Snippet in the search bar. Then, select **HTML Snippet Pack** and click **Download** (see *Figure 14.2*):

Figure 14.2 – HTML Snippet Pack in the Manage Extensions tool

We will receive a notification, which means we need to close and open (restart) Visual Studio to complete the installation. After opening Visual Studio again, it will ask for confirmation to install the extension. Click on **Modify** (see *Figure 14.3*):

Figure 14.3 – Installing HTML Snippet Pack

Then, we will see a progress bar, indicating that the installation is in progress. Finally, we will see a successful message:

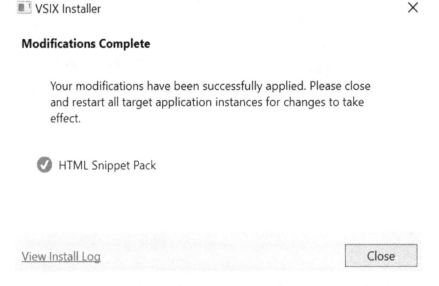

Figure 14.4 – Installation of HTML Snippet Pack extension completed

Click **Close** and continue to Visual Studio.

Using HTML Snippet Pack

We need to navigate to an HTML file to use the HTML Snippet Pack extension, so let's go to **SPAProject** | **ClientApp** | **public** | `index.html`.

Within this file, we can write the word `article` in the body element (see *Figure 14.5*):

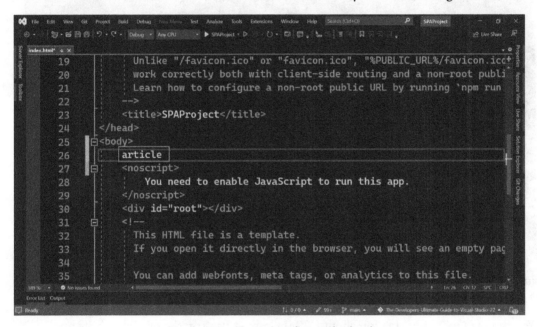

Figure 14.5 – Typing article in index.html

After writing the word `article`, press *tab* on your keyboard to easily generate the HTML element for an article in this part of the code. In *Figure 14.6*, we can see the article element generated automatically:

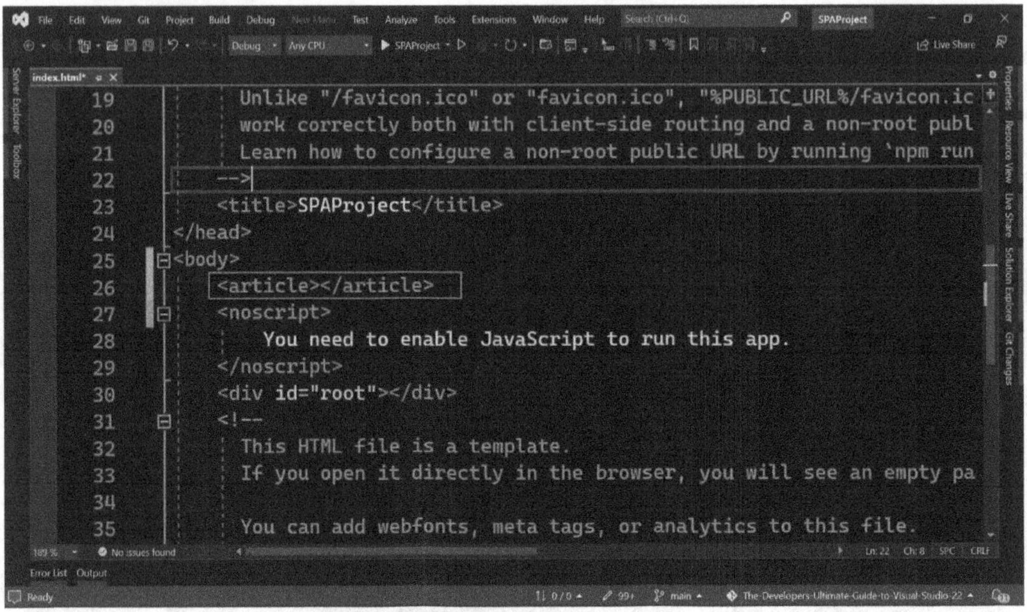

Figure 14.6 – The article generated using a code snippet

Just as we generated the `article` element, we can easily create elements for `li`, `ul`, `img`, `input`, and almost all the existing elements in the HTML standard.

You can read more information about HTML Snippet Pack on the official website and repository at `https://github.com/madskristensen/HtmlSnippetPack`.

Let's now review a different extension to analyze our code.

Cleaning up code with CodeMaid

CodeMaid is an amazing extension that helps us to simplify and clean up code. It is free and compatible with C#, C++, F#, VB, PHP, PowerShell, R, JSON, XAML, XML, ASP, HTML, CSS, LESS, SCSS, JavaScript, and TypeScript.

You can read more information about CodeMaid on the official website: `https://www.codemaid.net/`.

Let's install CodeMaid and see how we can use it.

Installing CodeMaid in Visual Studio 2022

To install CodeMaid, navigate to **Extensions | Manage Extensions**, type `codemaid` in the search bar, and then select **CodeMaid VS2022** (see *Figure 14.7*):

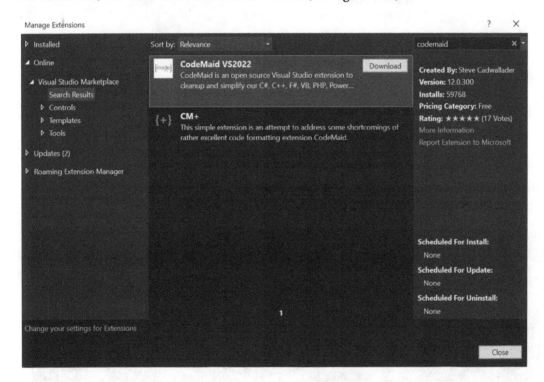

Figure 14.7 – Installing CodeMaid from the Manage Extensions tool

After searching `codemaid`, click **Download** and get the extension. We will get a message confirming that the next time we open Visual Studio, the extension will be installed. Close and open Visual Studio with `SPAProject` again. You can click on **Modify** and then wait for the installation to complete:

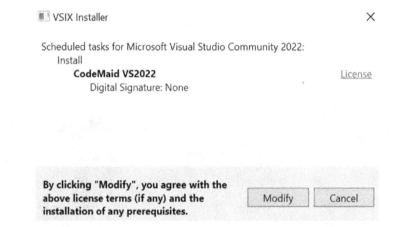

Figure 14.8 – The extension installer for CodeMaid in Visual Studio 2022

Now, you will see a new option in the **Extensions** menu that contains all the functionalities and configurations related to CodeMaid (see *Figure 14.9*):

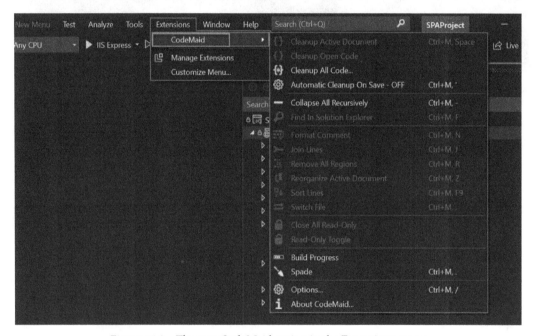

Figure 14.9 – The new CodeMaid option in the Extensions menu

Using CodeMaid

In this case, there is no active document, and therefore, many options are disabled, but we can use the **Cleanup All Code…** option to perform a cleanup on the whole project, using the default settings in CodeMaid. We will get a confirmation message before starting the process. Click on **Yes** to continue (see *Figure 14.10*):

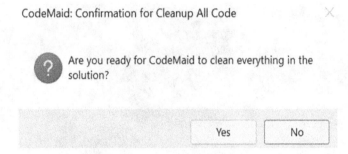

Figure 14.10 – Confirmation to clean up the project using CodeMaid

After confirming, CodeMaid will analyze each file at a time, look for white spaces and empty lines, and sort lines, among other things. You will see a progress bar and the current file being processed (see *Figure 14.11*):

```json
{
  "iisSettings": {
    "windowsAuthentication": false,
    "anonymousAuthentication": true,
    "iisExpress": {
      "applicationUrl": "http://localhost:29609",
      "sslPort": 44343
    }
  },
  "profiles": {
    "SPAProject": {
      "commandName": "
      "launchBrowser": true,
      "applicationUrl": "https://localhost:7132;http://localhost:5132",
      "environmentVariables": {
        "ASPNETCORE_ENVIRONMENT": "Development",
        "ASPNETCORE_HOSTINGSTARTUPASSEMBLIES": "Microsoft.AspNetCore.SpaProxy"
```

Figure 14.11 – A cleanup in process using CodeMaid

After completing the cleanup, all the files will be saved. We can open a file to see the changes, but using the Git integration in Visual Studio, we can see the differences easily. This works only if you already have the project connected with a Git repo. You can go to the **Git Changes** tab and open `Program.cs` to see the changes. An example of the changes is shown in *Figure 14.12*:

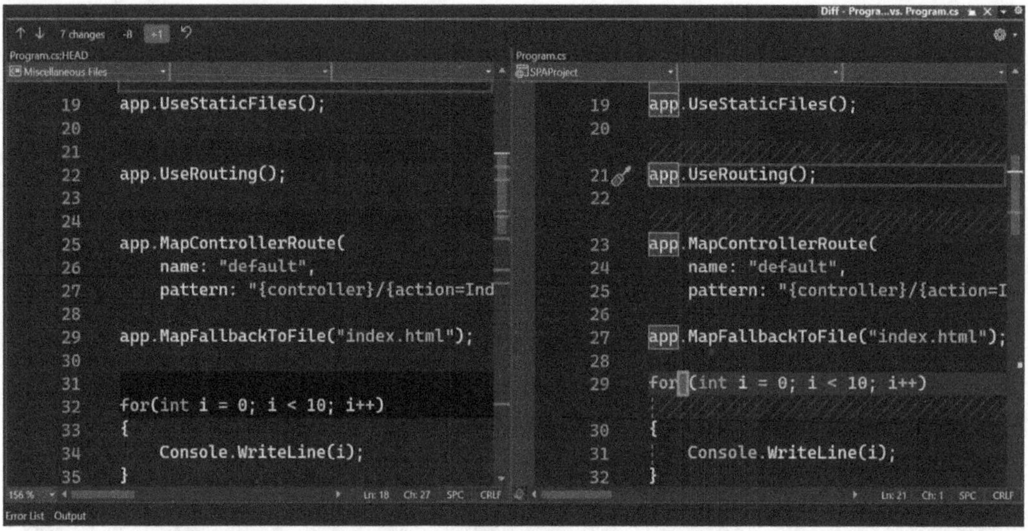

Figure 14.12 – A clean up in process using CodeMaid

In `Program.cs`, we can see two lines with some spaces in the `if` statement that CodeMaid removed.

> **Important Note**
> Reducing lines in code means reducing the size of a file. Blank lines and white spaces make code difficult to read and increase a project size when published.

CodeMaid has some options that we can turn on or turn off, depending on our needs. Navigate to **Extensions | CodeMaid | Options** and select the **Remove** section to choose scenarios where CodeMaid can remove code (see *Figure 14.13*):

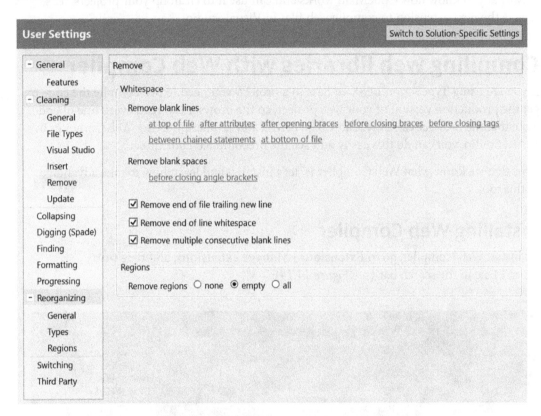

Figure 14.13 – User Settings for CodeMaid

CodeMaid also has many options related to removing blank lines and blank spaces. By default, all the options are enabled, but you can click on the options that you want to disable to set up CodeMaid according to the guidelines in your project.

You can read more information about CodeMaid on the official website: `https://www.codemaid.net/`.

Now that you know how CodeMaid works and can use it to clean up your projects, let's review the next extension to compile web files in Visual Studio.

Compiling web libraries with Web Compiler

If you are using TypeScript, LESS, or Sass in a project, you need to pre-compile the code to get the production version of your web project, so the browser can read every line of code. Note that the browser can only read CSS, HTML, and JavaScript. Using Web Compiler in Visual Studio, you can do this easily and see the precompiled code directly.

Now that we know what Web Compiler is, let's install it and learn how to take advantage of this tool.

Installing Web Compiler

To install Web Compiler, go to **Extensions | Manage extensions**, and type web `compiler` in the search bar (see *Figure 14.14*):

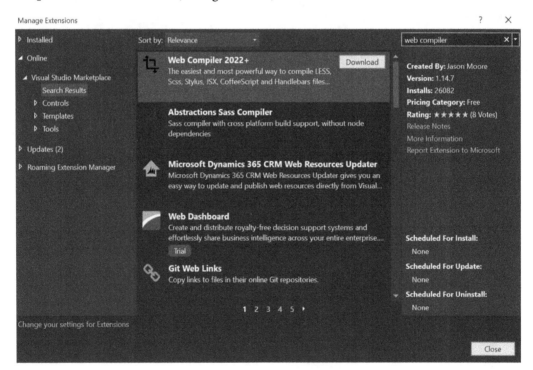

Figure 14.14 – Searching for Web Compiler in Manage Extensions

We can now follow the same steps that we did in the *Adding HTML Snippet Pack* section to complete the installation. After closing and then opening Visual Studio, you will see that the installation of this new extension is completed. Now, we will be able to use it in our `SPAProject`.

Using Web Compiler

To use Web Compiler, we can navigate and select any JavaScript file in the project, and we will see a new option in the menu after right-clicking. For example, let's select **ClientApp** | `aspnetcore-react.js` (see *Figure 14.15*):

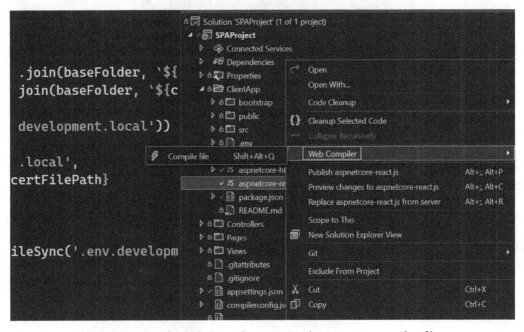

Figure 14.15 – The Web Compiler option in the project to compile a file

Using the **Web Compiler | Compile file** option, we can compile the file and generate a new version using **ECMAScript 2009 (ES5)**, which is a JavaScript specification that allows us to support old browser versions. See the file generated in *Figure 14.16*:

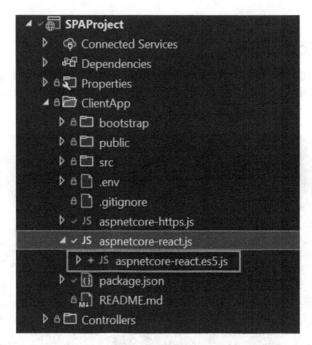

Figure 14.16 – aspnetcore-react.es5.js, generated by Web Compiler

The `aspnetcore.es5.js` file contains the same logic as `aspnetcore-react.js`, but it uses different syntaxes.

The following code represents the first 20 lines of the `aspnetcore-react.js` file:

```
// This script configures the .env.development.local file with
additional environment variables to configure HTTPS using the
ASP.NET Core
// development certificate in the webpack development proxy.

const fs = require('fs');
const path = require('path');

const baseFolder =
    process.env.APPDATA !== undefined && process.env.APPDATA
!== ''
        ? `${process.env.APPDATA}/ASP.NET/https`
```

```
           : `${process.env.HOME}/.aspnet/https`;

const certificateArg = process.argv.map(arg => arg.match(/--
name=(?<value>.+)/i)).filter(Boolean)[0];
const certificateName = certificateArg ? certificateArg.groups.
value : process.env.npm_package_name;

if (!certificateName) {
    console.error('Invalid certificate name. Run this script
in the context of an npm/yarn script or pass --name=<<app>>
explicitly.')
    process.exit(-1);
}


const certFilePath = path.join(baseFolder, `${certificateName}.
pem`);
```

The aspnetcore-react.js file contains some logic related to the interaction between the asp.net application in the backend and the React application. This file uses the last features in JavaScript, so Web Compiler needs to compile the code to an old version to support the old browser and increase the compatibility with other libraries.

The following code represents the first 20 lines of the aspnetcore-react.es5.js file:

```
// This script configures the .env.development.local file with
additional environment variables to configure HTTPS using the
ASP.NET Core
// development certificate in the webpack development proxy.

'use strict';

var fs = require('fs');
var path = require('path');

var baseFolder = process.env.APPDATA !== undefined && process.
env.APPDATA !== '' ? process.env.APPDATA + '/ASP.NET/https' :
process.env.HOME + '/.aspnet/https';

var certificateArg = process.argv.map(function (arg) {
    return arg.match(/--name=(?<value>.+)/i);
```

```
}).filter(Boolean)[0];
var certificateName = certificateArg ? certificateArg.groups.
value : process.env.npm_package_name;

if (!certificateName) {
    console.error('Invalid certificate name. Run this script
in the context of an npm/yarn script or pass --name=<<app>>
explicitly.');
    process.exit(-1);
}
```

Each file added to Web Compiler will be included in a file called `compilerconfig.json`. This file is associated with the compilation process with Visual Studio. This means that we can build and publish the project normally, and the files associated will be generated automatically.

This is an example of the `compilerconfig.json` file after using it in the `aspnetcore-react.js` file:

```
[
    {
        "outputFile": "ClientApp/aspnetcore-https.es5.js",
        "inputFile": "ClientApp/aspnetcore-https.js"
    },
    {
        "outputFile": "ClientApp/aspnetcore-react.es5.js",
        "inputFile": "ClientApp/aspnetcore-react.js"
    }
]
```

Each configuration is a JSON object that contains two properties – `inputFile` is the location of the source file to compile, and `outputFile` has the location of the file generated by Web Compiler.

> **Important Note**
> You can use *Shift + Alt + Y* to compile all the files included in
> `compilerconfig.json`.

For more information, you can read the documentation on GitHub: `https://github.com/failwyn/WebCompiler`.

Now you know how to use Web Compiler and how to transform JavaScript files to use ES5. Let's see another extension that extends the functionalities in our editor and allows us to see some guides and easily distinguish white and blank spaces.

Identifying white spaces with Indent Guides

In *Chapter 13*, *Extensions in Visual Studio*, we reviewed some extensions in Visual Studio, including Color Preview. With these extensions, we learned how the text editor in Visual Studio can be extended to improve our experience and provide more tools for some technologies and scenarios. Indent Guides is another example of this type of extension that extends the text editor in Visual Studio.

Indent Guides is a simple but useful extension that helps us identify extra white spaces and indentations in the structure of code.

Installing Indent Guides

To install Indent Guides, go to **Extensions | Manage Extensions** and type `Indent Guides` in the search bar (see *Figure 14.17*):

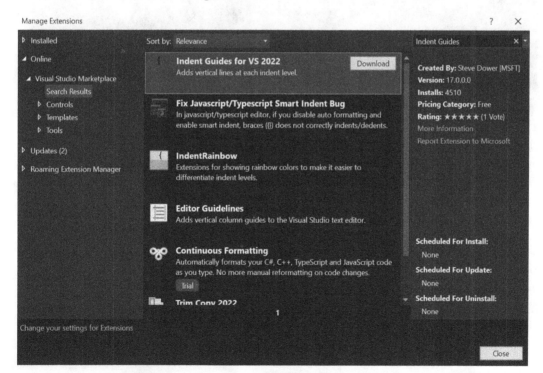

Figure 14.17 – Indent Guides in Manage Extensions

Click on **Download** to get this extension and then close and open Visual Studio to install it. At this point, follow the steps from the *Adding HTML Snippet Pack* section to complete the installation. Once you have Visual Studio running again and the installation is completed, you can open the `index.html` file and see new guides that show us white spaces and tabs between text and the elements in the text editor (see *Figure 14.18*):

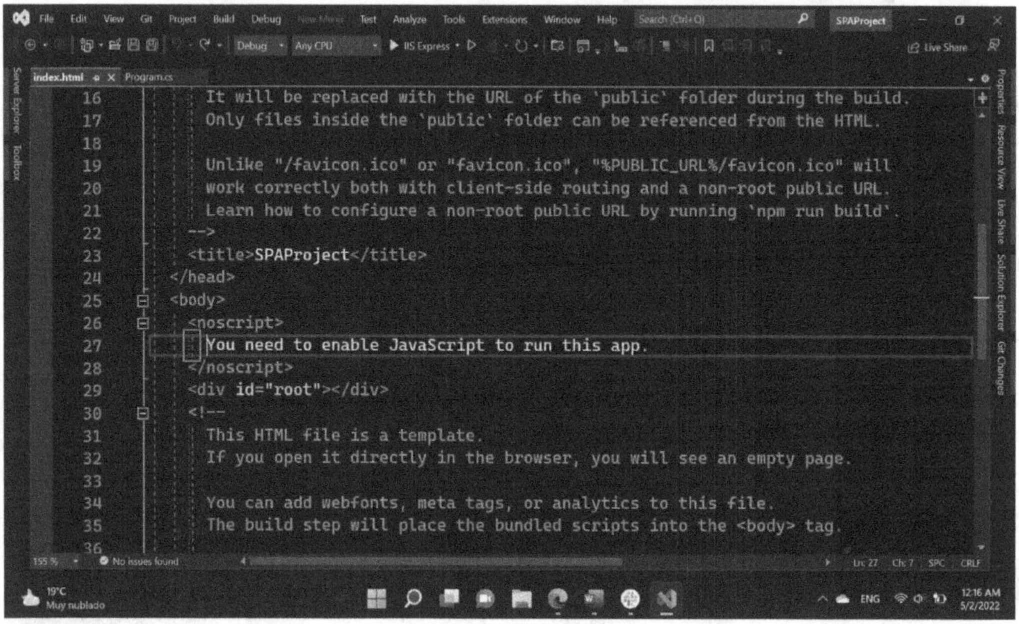

Figure 14.18 – Guides in the Visual Studio text editor

This tool is amazing for improving the formatting in our files and is compatible with all the programming languages supported by Visual Studio. There are some additional options that we can adjust to fit our preferences. Navigate to **Tools | Options | Indent Guides**. There, we will find many options to change the appearance, behavior, and highlights and set a quick start or default configuration.

Summary

Visual Studio has a marketplace with many extensions that we can access using the Visual Studio **Manage Extensions** option. We can find many extensions related to code snippets in the Studio Marketplace and use HTML Snippet Pack to improve our productivity while coding in HTML files.

You now know how to use CodeMaid to clean up code and improve code quality in your projects. You can easily remove empty lines and white spaces and create a custom validation format for all the files in your project. You also learned how to install Web Compiler to compile and transform web files such as JavaScript files or libraries such as LESS and Sass into generic code that a browser can easily read. Finally, you explored the Indent Guides extension, which shows us white and blank spaces in code to identify how to improve the format and structure of files.

In *Chapter 15*, *Learning Keyboard Shortcuts*, you will review the most important shortcuts included in Visual Studio by default. With this knowledge, you will use a combination of some keyboard keys to perform common actions in Visual Studio.

15
Learning Keyboard Shortcuts

During this book, we have analyzed some shortcuts that help us improve our productivity, using a combination of some keys to perform common actions in the IDE and source code. Visual Studio includes some useful shortcuts by default, but we also have the possibility of creating our own shortcuts, depending on our needs or common operations that we need to perform daily.

In this chapter, we will provide a summary of the most important shortcuts included by default in Visual Studio 2022 and a section that shows how to create our own.

We will review the following topics in this chapter:

- The most-used shortcuts for use in source code
- The most common shortcuts for use in the IDE
- Creating custom shortcuts

When working with Visual Studio, it is extremely important that we know the keyboard shortcuts so that we can carry out operations quickly. This will prevent you from wasting time on repetitive tasks, such as formatting a complete source code file or renaming a member of a class.

It is important you put these keyboard shortcuts into practice and begin to use them little by little, even if you execute them slowly at the beginning. You will see that with time, you will execute them automatically, without resorting to any visual aid.

Let's learn how to improve our productivity using shortcuts.

Technical requirements

To use shortcuts in this project, you must have previously installed Visual Studio 2022 with the web development workload installed, as shown in *Chapter 1, Getting Started with Visual Studio*. It's also important to have the SPA base project created in *Chapter 4, Projects and Templates*.

It is important to note that there are different keyboard mapping schemes that can completely modify the shortcut keys. In addition, extensions such as ReSharper can also modify keyboard shortcuts. For correct execution of all the shortcuts shown in this chapter, you must have a **Default** keyboard mapping scheme, which can be selected from the **Tools | Options | Environment | Keyboard** menu.

Let's enter the world of shortcuts, which will undoubtedly make you become a more efficient and productive programmer.

Shortcuts for use in source code

Working with source code involves working with thousands of lines of code, which can be a headache for even the most experienced developer. That is why the powerful search tools included in Visual Studio are an excellent way to search and navigate throughout the lines of code.

Similarly, code editing and refactoring tasks are ongoing in projects, not to mention debugging and testing in large projects.

This is why working with shortcuts that give you instant access to these tools should be a priority in your career as a developer. Let's see what these keyboard shortcuts are.

Shortcuts for searching and navigating through source code

Finding members of a class quickly may seem like a simple task when working on a one-class project, but you may not think so if you work with projects that contain hundreds of classes or even several projects within the same solution. It is during these moments that the following keyboard shortcuts become a great help.

Let's take a look at the shortcuts that will help us perform quick search and navigation operations in Visual Studio:

- *Ctrl + Q*: Displays **Visual Studio Search**
- *Ctrl + T*: Displays the **Go To All** tool
- *Ctrl + -*: Navigates backward between opened documents in the current session
- *Ctrl + Shift + -*: Navigates forward between opened documents in the current session
- *F12*: Navigates to a class definition
- *Alt + F12*: Allows you to view and edit the code of a class in a pop-up window, from the code file you are writing.
- *Ctrl + F12*: Navigates to the implementation of a member in a class
- *Ctrl + Shift + F12*: Moves to the next error line when the error list window is open and more than one error is listed
- *F8*: Navigates forward in the list of results of the current window
- *Shift + F8*: Navigates backward in the list of results of the current window

The shortcuts discussed here allow you to quickly navigate between files, members, results, and implementations without taking your hands away from the keyboard. Now, let's look at the most common shortcuts for editing and refactoring.

> **Important Note**
> You can check out all the most common shortcuts in Visual Studio at the following link: `https://docs.microsoft.com/visualstudio/ide/default-keyboard-shortcuts-in-visual-studio?view=vs-2022`. We encourage you to download the file, print it, and keep it near you for quick reference to these shortcuts.

Shortcuts for editing and refactoring

The commands that we will see in this subsection correspond to those that allow you to apply changes directly to source code. Among the most common operations are renaming members, commenting on lines of code, and moving lines up and down. Let's see the list of shortcuts that will make writing code even easier:

- *Alt + Enter*: Displays quick actions
- *Ctrl + K, Ctrl + I*: Obtains information about a member of a class

- *Ctrl + K, Ctrl + C*: Comments multiple selected lines in the source code
- *Ctrl + K, Ctrl + U*: Uncomments multiple selected lines in the source code
- *Ctrl + Shift + L*: Deletes selected lines
- *Ctrl + Shift + V*: Displays and allows pasting the contents from the buffer ring (it refers to the history of elements that have been previously copied)
- *Ctrl + F*: Finds a specific text in the code
- *Ctrl + A*: Selects all the lines in the current file
- *Ctrl + S*: Saves the pending changes in the current file
- *Ctrl + Shift + S*: Saves the pending changes in all the opened files
- *Ctrl + Shift + .*: Zooms into the current file
- *Ctrl + Shift + ,*: Zooms out from the current file
- *Ctrl + Up*: Moves selected lines up in a code file
- *Ctrl + Down*: Moves selected lines down in a code file
- *Ctrl + K, Ctrl + D*: Applies the style rules to the entire document
- *Ctrl + K, Ctrl + F*: Applies the style rules only in the selected lines in the document
- *Ctrl + K, Ctrl + S*: Used to encapsulate the code between common clauses, such as cycles (`while`, `for`, and so on), control statements (`if`, `switch`, and so on), or code regions (`#region`).
- *Ctrl + R, Ctrl + R*: Renames a member
- *Ctrl + R, Ctrl + E*: Creates a property for a field of a class
- *Ctrl + R, Ctrl + G*: Removes unused `using` statements in a class and sorts them alphabetically
- *Ctrl + R, Ctrl + M*: Creates a method from the selected code

Now that we have learned about the main shortcuts for editing and refactoring code, let's take a look at those that will help us to optimize depuration and testing tasks.

Shortcuts for debugging and testing

Debugging and code execution is one of the most constant tasks we will perform while working with Visual Studio. Therefore, it is important to know the keyboard shortcuts that can help us to execute these tasks quickly. That is why, in this subsection, we will mention the most important shortcuts focused on these tasks:

- *F5*: Starts the application in debug mode
- *Ctrl + F5*: Starts the application without debug mode
- *Shift + 5*: Stops the application when it's running
- *Ctrl + Shift + F5*: Stops the application execution, rebuilds the project, and creates a new debugging session
- *F9*: Places or removes a breakpoint
- *F10*: Skips the execution of code when debugging
- *F11*: Debugs source code line by line
- *Shift + F11*: Steps out of the execution of the method
- *Ctrl + R, Ctrl + A*: Starts unit test execution in debug mode
- *Ctrl + R, A*: Starts unit test execution without debug mode

This concludes the list of shortcuts that can help us to improve our time when working with source code. Let's now review the shortcuts that can help us perform quick actions in the IDE environment.

The most common shortcuts for use in the IDE

Knowing how to get around in the Visual Studio IDE through keyboard shortcuts is an important part of avoiding wasting time searching through menus to activate a specific panel. It is very common, for example, to close by mistake the **Solution Explorer** or **Properties** window and not know which menu contains the option to open them again. That is why, in this section, we will examine the shortcuts that will speed up the performance of these tasks:

- *Ctrl + [+ S*: If we have a file open, this shortcut allows us to quickly select it in the **Solution Explorer** window.

- *Ctrl + Alt + L*: Opens the **Solution Explorer** window.
- *Ctrl + Alt + O*: Opens the **Output** window.
- *Ctrl + \, E*: Opens the **Error List** window.
- *Ctrl + \, Ctrl + M*: Opens the **Team Explorer** window.
- *Ctrl + Alt + B*: Opens the **Breakpoints** window.
- *F4*: Opens the **Properties** window.
- *Alt + F6*: Allows you to scroll back between windows on the panels that are open.
- *Shift + Alt + F6*: Allows you to scroll forward between windows on the panels that are open.
- *Shift + Esc*: Allows you to close the current tool window.
- *Ctrl + Alt + Pg Up*: Allows scrolling up between open documents even when it is not the same session.
- *Ctrl + Alt + Pg Dn*: Allows scrolling down between open documents even when it is not the same session.
- *Ctrl + Tab*: Displays a special window with the open documents and selects the most recent one.
- *Ctrl + Shift + Tab*: Displays a special window with the open documents and selects the least recent one.
- *Shift + Alt + Enter*: Allows you to place the Visual Studio environment at full screen, allowing you to concentrate on the current document. Use the same shortcut to get out of full-screen mode.
- *Ctrl + K + K*: Creates a bookmark in the line where we are positioned. If there is already a bookmark on that line, it will be deleted.

> **Important Note**
>
> Bookmarks are a feature of Visual Studio that allow you to mark lines in your code to quickly return to them. You can find more information about them at the following link: `https://docs.microsoft.com/en-us/visualstudio/ide/setting-bookmarks-in-code?view=vs-2022`.

- *Ctrl + K + N*: Allows you to scroll forward through the different bookmarks that are part of a project.

- *Ctrl + K + P*: Allows you to scroll backward through the different bookmarks that are part of a project.

At this point, you have learned the most common and useful shortcuts in Visual Studio, but there is a way to create our own shortcuts to adapt Visual Studio for our needs. Let's analyze step by step how to create custom shortcuts.

Creating custom shortcuts

We can create our own shortcuts for specific actions in Visual Studio, and there are several options available to customize the current shortcuts.

You can navigate to **Tools | Options | Environment | Keyboard** to see all the current shortcuts in Visual Studio (as shown in *Figure 15.1*):

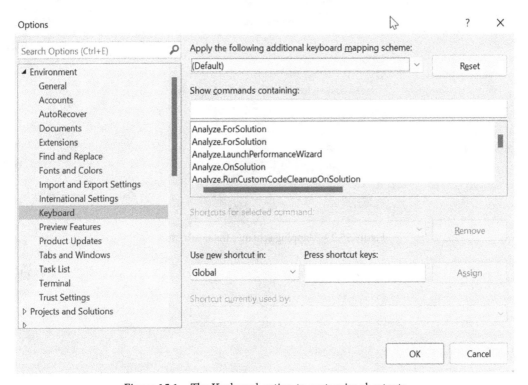

Figure 15.1 – The Keyboard option to customize shortcuts

You will find all the current shortcuts in Visual Studio for all the functionalities and a scheme for the shortcuts, where you can set them up, depending on the context. By default, Visual Studio includes different keyboard schemes with different keyboard shortcut configurations. (see *Figure 15.2*):

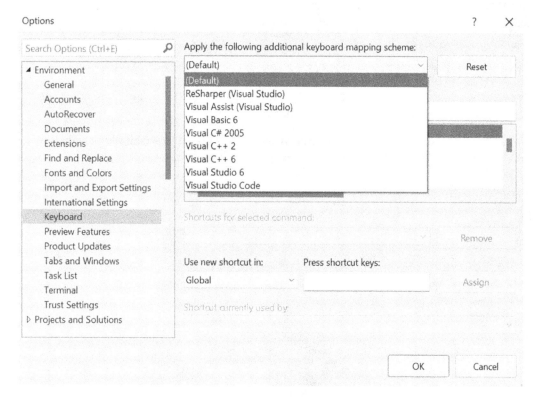

Figure 15.2 – Mapping schemes for shortcuts

To create a new shortcut, select (**Default**) in the **Apply the following additional keyboard mapping scheme** option, and then select the **Analyze. RunDefaultCodeCleanUpOnSolution** command. This command executes a process to clean up code to improve the format and remove unnecessary code. Finally, you can assign the shortcut for this command by adding the *Ctrl + Alt + Y* combination in the **Press shortcut keys** option (see *Figure 15.3*):

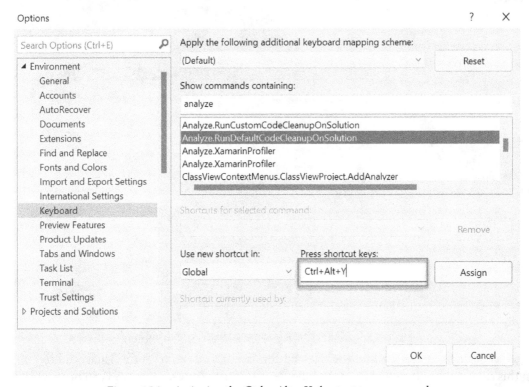

Figure 15.3 – Assigning the Ctrl + Alt + Y shortcut to a command

Now, click **Assign** and then **OK** to confirm and add this new shortcut in Visual Studio, and perform the cleanup process by default in the solution quickly (see *Figure 15.4*):

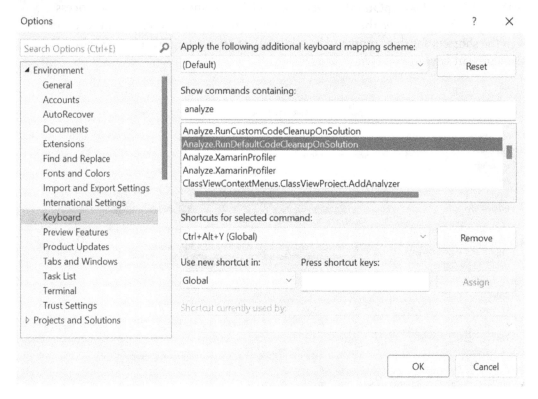

Figure 15.4 – The shortcut is assigned to the command

After assigning the shortcuts for the selected command, you can try it in Visual Studio. In this case, the **Analyze.RunDefaultCodeCleanUpOnSolution** command can be used globally, so you only need to have the SPAProject open, and after pressing *Ctrl + Alt + Y*, Visual Studio will perform a cleanup on the solution.

> **Important Note**
>
> You can override shortcuts included by default in Visual Studio. However, this is not the best practice, since you are altering the normal behavior in Visual Studio, and it could be difficult to work in other environments.

Summary

In this chapter, we learned about shortcuts in Visual Studio, and we reviewed all the useful shortcuts that we can use while coding or performing an action with tools or functionalities.

With this knowledge on how to use shortcuts in your daily work, you will depend less and less on the keyboard to execute actions within Visual Studio, which will allow you to become an efficient developer.

We also learned how to create our own shortcuts and automate common tasks in our projects using a key combination.

It has been a long journey from where we started – analyzing what Visual Studio is in *Chapter 1*, *Getting Started with Visual Studio 2022*, what the available versions are, and how to install it in our local environment – to the current chapter where we learned about shortcuts.

At this point, we must congratulate you for having reached the end of the book. We hope you enjoyed reading it as much as we enjoyed writing it. The next step is to get down to work and apply the knowledge you have acquired in your day-to-day work as a developer. Happy coding.

Index

`Packt.com`

Subscribe to our online digital library for full access to over 7,000 books and videos, as well as industry leading tools to help you plan your personal development and advance your career. For more information, please visit our website.

Why subscribe?

- Spend less time learning and more time coding with practical eBooks and Videos from over 4,000 industry professionals

- Improve your learning with Skill Plans built especially for you

- Get a free eBook or video every month

- Fully searchable for easy access to vital information

- Copy and paste, print, and bookmark content

Did you know that Packt offers eBook versions of every book published, with PDF and ePub files available? You can upgrade to the eBook version at `packt.com` and as a print book customer, you are entitled to a discount on the eBook copy. Get in touch with us at `customercare@packtpub.com` for more details.

At `www.packt.com`, you can also read a collection of free technical articles, sign up for a range of free newsletters, and receive exclusive discounts and offers on Packt books and eBooks.

Other Books You May Enjoy

If you enjoyed this book, you may be interested in these other books by Packt:

Enterprise Application Development with C# 10 and .NET 6

Suneel Kumar Kunani | Arun Kumar Tamirisa | Bhupesh Guptha Muthiyalu | Ravindra Akella

ISBN: 978-1-80323-297-3

- Design enterprise apps by making the most of the latest features of .NET 6
- Discover different layers of an app, such as the data layer, API layer, and web layer
- Explore end-to-end architecture by implementing an enterprise web app using .NET and C# 10 and deploying it on Azure
- Focus on the core concepts of web application development and implement them in .NET 6

Customizing ASP.NET Core 6.0

Jürgen Gutsch

ISBN: 978-1-80323-360-4

- Explore various application configurations and providers in ASP.NET Core 6
- Enable and work with caches to improve the performance of your application
- Understand dependency injection in .NET and learn how to add third-party DI containers
- Discover the concept of middleware and write your middleware for ASP.NET Core apps
- Create various API output formats in your API-driven projects

Hi!

We're Miguel and Hector, the authors of Hands-On Visual Studio 2022. We really hope you enjoyed reading this book and found it useful for increasing your productivity and efficiency in Visual Studio.

It would really help us (and other potential readers!) if you could leave a review on Amazon sharing your thoughts on Hands-On Visual Studio 2022 here.

Go to the link below or scan the QR code to leave your review:

https://packt.link/r/1801810540

Your review will help us to understand what's worked well in this book, and what could be improved upon for future editions, so it really is appreciated.

Best Wishes,

Miguel Angel Teheran Garcia

Hector Uriel Perez Rojas

Packt is searching for authors like you

If you're interested in becoming an author for Packt, please visit authors. packtpub.com and apply today. We have worked with thousands of developers and tech professionals, just like you, to help them share their insight with the global tech community. You can make a general application, apply for a specific hot topic that we are recruiting an author for, or submit your own idea.

www.ingramcontent.com/pod-product-compliance
Lightning Source LLC
Chambersburg PA
CBHW062056050326
40690CB00016B/3105